PENGUIN BOOKS

IMPERIAL AMBITIONS

Noam Chomsky is the author of numerous bestselling political books, including, most recently, *9-11*, *Understanding Power*, *Middle East Illusions*, *Pirates and Emperors, Old and New* and *Hegemony or Survival*. He is a professor in the Department of Linguistics and Philosophy at MIT, and is widely credited with having revolutionized modern linguistics.

David Barsamian is the founder and director of the widely syndicated weekly show *Alternative Radio*, and is the author of several books of interviews with leading political thinkers, including Arundhati Roy, Howard Zinn, Tariq Ali and especially Noam Chomsky.

NOAM CHOMSKY

IMPERIAL AMBITIONS

CONVERSATIONS
WITH NOAM CHOMSKY
ON THE POST-9/11 WORLD

INTERVIEWS WITH

DAVID BARSAMIAN

PENGUIN BOOKS

PENGUIN BOOKS

Published by the Penguin Group
Penguin Books Ltd, 80 Strand, London WC2R 0RL, England
Penguin Group (USA) Inc., 375 Hudson Street, New York, New York 10014, USA
Penguin Group (Canada), 90 Eglinton Avenue East, Suite 700, Toronto, Ontario, Canada M4P 2Y3
(a division of Pearson Penguin Canada Inc.)
Penguin Ireland, 25 St Stephen's Green, Dublin 2, Ireland (a division of Penguin Books Ltd)
Penguin Group (Australia), 250 Camberwell Road, Camberwell,
Victoria 3124, Australia (a division of Pearson Australia Group Pty Ltd)
Penguin Books India Pvt Ltd, 11 Community Centre,
Panchsheel Park, New Delhi – 110 017, India
Penguin Group (NZ), cnr Airborne and Rosedale Roads, Albany,
Auckland 1310, New Zealand (a division of Pearson New Zealand Ltd)
Penguin Books (South Africa) (Pty) Ltd, 24 Sturdee Avenue,
Rosebank, Johannesburg 2196, South Africa

Penguin Books Ltd, Registered Offices: 80 Strand, London WC2R 0RL, England

www.penguin.com

First published in the United States of America by Metropolitan Books,
Henry Holt and Company, LLC 2005
First published in Great Britain by Hamish Hamilton 2005
Published in Penguin Books 2006

1

Printed in England by Clays Ltd, St Ives plc

ISBN-13: 978-0-141-02692-3
ISBN-10: 0-141-02692-8

CONTENTS

INTRODUCTION

I'm frequently asked, What's it like to interview Noam Chomsky? In more than twenty years of working with him, I've learned several things. One is, be prepared and put your questions in some order of priority. Another is, listen carefully, because you never know which way the conversation will go.

Chomsky's soft voice masks a torrent of information and analysis. He has an extraordinary power to distill and synthesize reams of information. And he misses nothing. In one interview he referred to the 1988 shooting down of a civilian Iranian airliner by the USS *Vincennes*. I was flabbergasted to learn that his source was *Proceedings*, the journal of the U.S. Naval Institute.

I began Alternative Radio with a series of Chomsky interviews in 1986, and we have never stopped talking since. The interviews in this collection were mostly conducted in Chomsky's office at MIT. The interview questions were unrehearsed. For this book we have edited the transcripts, expanded on our discussions, and added notes.

So what's it like to interview Chomsky? It's to be in the presence of someone who insists it's not so complicated to understand the truth or to know how to act. Someone who defines and embodies what intellectuals *should* be. Who excoriates those who genuflect before power and denounce others while avoiding their own responsibility.

Chomsky sets the compass headings and describes the topography. It is up to us to navigate the terrain. It is my hope that the conversations in this book will spark thought, discussion, and, most of all, activism.

Special thanks to Anthony Arnove, comrade, friend, and editor par excellence; Sara Bershtel, publisher and editor par excellence; Elaine Bernard for her generosity; Greg Gigg for his suggestions; KGNU community radio; David Peterson, Chris Peterson, and Dale Wertz for their research assistance; Bev Stohl for accommodating my numerous requests; Martin Voelker for his technical support and friendship; and to Noam Chomsky for his solidarity, patience, and great sense of humor.

Sections of some of these interviews have appeared in different forms in *International Socialist Review, Monthly Review, The Progressive, The Sun,* and Z.

DAVID BARSAMIAN
Boulder, Colorado, July 2005

IMPERIAL
AMBITIONS

IMPERIAL AMBITIONS

CAMBRIDGE, MASSACHUSETTS (MARCH 22, 2003)

What are the regional implications of the U.S. invasion and occupation of Iraq?

I think not only the region but the world in general correctly perceives the U.S. invasion as a test case, an effort to establish a new norm for the use of military force. This new norm was articulated in general terms by the White House in September 2002 when it announced the new *National Security Strategy of the United States of America.*[1] The report proposed a somewhat novel and unusually extreme doctrine on the use of force in the world, and it's not accidental that the drumbeat for war in Iraq coincided with the report's release.

The new doctrine was not one of *preemptive* war, which arguably falls within some stretched interpretation of the UN Charter, but rather a doctrine that doesn't begin to have any grounds in international law, namely, *preventive* war. That is, the United States will rule the world by force, and if there is any challenge to its domination—whether it is perceived in the distance, invented, imagined, or whatever—then the United States will have the right to destroy that challenge before it becomes a threat. That's preventive war, not preemptive war.

To establish a new norm, you have to do something. Of course, not every state has the capacity to create what is called a new norm. So if India invades Pakistan to put an end to monstrous atrocities, that's not a norm. But if the United States bombs Serbia on dubious grounds, that's a norm. That's what power means.

The easiest way to establish a new norm, such as the right of preventive war, is to select a completely defenseless target, which can be easily overwhelmed by the most massive military force in human history. However, in order to do that credibly, at least in the eyes of your own population, you have to frighten people. So the defenseless target has to be characterized as an awesome threat to survival that was responsible for September 11 and is about to attack us again, and so on. And this was indeed done in the case of Iraq. In a really spectacular propaganda achievement, which will no doubt go down in history, Washington undertook a massive effort to convince Americans, alone in the world, that Saddam Hussein was not only a monster but also a threat to our existence. And it substantially succeeded. Half the U.S. population believes that Saddam Hussein was "personally involved" in the September 11, 2001, attacks.[2]

So all this falls together. The doctrine is pronounced, the norm is established in a very easy case, the population is driven into a panic and, alone in the world, believes the fantastic threats to its existence, and is therefore willing to support military force in self-defense. And if you believe all of this, then it really is self-defense to invade Iraq, even though in reality the war is a textbook example of aggression, with the purpose of extending the scope for further aggression. Once the easy case is handled, you can move on to harder cases.

Much of the world is overwhelmingly opposed to the war because they see that this is not just about an attack on Iraq. Many people correctly perceive it exactly the way it's intended, as a firm statement that you had better watch out, you could be next. That's why the United

States is now regarded as the greatest threat to peace in the world by a large number of people, probably the vast majority of the population of the world. George Bush has succeeded within a year in converting the United States to a country that is greatly feared, disliked, and even hated.[3]

At the World Social Forum in Porto Alegre, Brazil, in February 2003, you described Bush and the people around him as "radical nationalists" engaging in "imperial violence."[4] Is this regime in Washington, D.C., substantively different from previous ones?

It is useful to have some historical perspective, so let's go to the opposite end of the political spectrum, about as far as you can get, the Kennedy liberals. In 1963, they announced a doctrine which is not very different from Bush's National Security Strategy. Dean Acheson, a respected elder statesman and a senior adviser to the Kennedy administration, delivered a lecture to the American Society of International Law in which he stated that no "legal issue" arises if the United States responds to any challenge to its "power, position, and prestige."[5] The timing of his statement is quite significant. He made it shortly after the 1962 Cuban missile crisis, which virtually drove the world to the edge of nuclear war. The

Cuban missile crisis was largely a result of a major campaign of international terrorism aimed at overthrowing Castro—what's now called *regime change*, which spurred Cuba to bring in Russian missiles as a defensive measure.

Acheson argued that the United States had the right of preventive war against a mere challenge to our position and prestige, not even a threat to our existence. His wording, in fact, is even more extreme than that of the Bush doctrine. On the other hand, to put it in perspective, this was a proclamation by Dean Acheson to the American Society of International Law; it wasn't an official statement of policy. The National Security Strategy document is a formal statement of policy, not just a statement by a high official, and it is unusual in its brazenness.

A slogan that we have all heard at peace rallies is "No Blood for Oil." The whole issue of oil is often referred to as the driving force behind the U.S. invasion and occupation of Iraq. How central is oil to U.S. strategy?

It's undoubtedly central. I don't think any sane person doubts that. The Gulf region has been the main energy-producing region of the world since the Second World War and is expected to be so for at least another generation. The Persian Gulf is a huge source of strategic power and material wealth. And Iraq is absolutely central to it. Iraq has the second largest oil reserves in the world, and Iraqi oil is very easily accessible and cheap. If you control Iraq, you are in a very strong position to determine the price and production levels (not too high, not too low) to undermine OPEC (the Organization of Petroleum Exporting Countries), and to throw your weight around throughout the world. This has nothing in particular to do with *access* to the oil for import into the United States. It's about *control* of the oil.

If Iraq were somewhere in central Africa, it wouldn't be chosen as a test case for the new doctrine of force, though this doesn't account for the specific timing of the current Iraq operation, because control over Middle East oil is a constant concern.

A 1945 State Department document on Saudi Arabian oil calls it "a stupendous source of strategic power, and one of the greatest material prizes in world history."[6] The United States imports quite a bit of its oil, about 15 percent, from Venezuela.[7] It also imports oil from Colombia and Nigeria. All three of these states are, from Washington's perspective, somewhat problematic right now, with Hugo Chávez in control in Venezuela, literally civil war in Colombia, and uprisings and strikes in Nigeria. What do you think about all of those factors?

All of this is very pertinent, and the regions you mention are where the United States actually intends to have access. In the Middle East, the United States wants control. But, at least according to intelligence projections, Washington intends to rely on what they regard as more stable Atlantic Basin resources, which means West Africa and the Western Hemisphere, areas that are more fully under U.S. control than is the Middle East, a difficult region. So disruption of one kind or another in those areas is a significant threat, and therefore another episode like Iraq is very likely, especially if the occupation works the way the civilian planners at the Pentagon hope. If it's an easy victory, with not too much fighting, and Washington can establish a new regime that it will call "democratic," they will be emboldened to undertake the next intervention.

You can think of several possibilities. One of them is the Andean region. The U.S. military has bases and soldiers all around the Andes now. Colombia and Venezuela, especially Venezuela, are both substantial oil producers, and there is more oil in Ecuador and Brazil. Another possibility is Iran.

Speaking of Iran, the Bush administration was advised by none other than, as Bush called him, the "man of peace," Ariel Sharon, to go after Iran "the day after" the United States finished with Iraq.[8] What about Iran, a designated "axis of evil" state and also a country that has significant oil reserves?

As far as Israel is concerned, Iraq has never been much of an issue. They consider it a kind of pushover. But Iran is a different story. Iran is a much more serious military and economic force. And for years Israel has been pressing the United States to take on Iran. Iran is too big for Israel to attack, so they want the big boys to do it.

And it's quite likely that this war may already be under way. A year ago, more than 10 percent of the Israeli air force was reported to be permanently based in eastern Turkey—at the huge U.S. military base there—and flying reconnaissance over the Iranian border. In addition, there are credible reports that the United States, Turkey, and Israel are attempting to stir up Azeri nationalist forces in northern Iran.[9] That is, an axis of U.S.-Turkish-Israeli power in the region opposed to Iran could ultimately lead to the split-up of Iran and maybe even to military attack, although a military attack will happen only if it's taken for granted that Iran would be basically defenseless. They're not going to invade anyone who they think can fight back.

With U.S. military forces in Afghanistan and in Iraq, as well as bases in Turkey, Iran is surrounded. The United States also has troops and bases throughout Central Asia to the north. Won't this encourage Iran to develop nuclear weapons, if they don't already have them, in self-defense?

Very likely. And the little serious evidence we have indicates that the Israeli bombing of Iraq's Osirak reactor in 1981 probably stimulated and may have initiated the Iraqi nuclear weapons development program.

But weren't they already engaged in it?

They were engaged in building a nuclear plant, but nobody knew its capacity. It was investigated on the ground after the bombing by a well-known nuclear physicist from Harvard, Richard Wilson. I believe he was head of Harvard's physics department at the time. Wilson published his analysis in a leading scientific journal, *Nature*.[10] He's an expert on this topic, and, according to Wilson, Osirak was a power plant. Other Iraqi exile sources have indicated that nothing much was going on; the Iraqis were toying with the idea of nuclear weapons before, but it was the bombing of Osirak that stimulated the nuclear weapons program.[11] You can't prove this, but that's what the evidence suggests.

What does the Iraq war and occupation mean for the Palestinians?

That's interesting to think about. One of the rules of journalism is that when you mention George Bush's name in an article, the headline has to speak of his "vision" and the article has to talk about his "dreams." Maybe there will be a photograph of him peering into the distance, right next to the article. It's become a journalistic convention. A lead story in the *Wall Street Journal* yesterday, had the words *vision* and *dream* about ten times.[12]

One of George Bush's dreams is to establish a Palestinian state somewhere, sometime, in some unspecified place—maybe in the Saudi desert. And we are supposed to praise that as a magnificent vision. But all this talk of Bush's vision and dream of a Palestinian state ignores completely that the United States would have to stop undermining the long-term efforts of the rest of the world, virtually without exception, to create some kind of a viable political settlement. For the last twenty-five to thirty years, the U.S. has been blocking any such settlement. The Bush administration has gone even further than others in blocking a solution, sometimes in such extreme ways that they weren't even reported. For example, in December 2002, the Bush administration reversed U.S. policy on Jerusalem. At least in principle, the United States had previously gone along with the 1968 Security Council resolution ordering Israel to revoke its annexation and occupation and settlement policies in East Jerusalem. But the Bush administration reversed that policy.[13] That's just one of many measures intended to undermine the possibility of any meaningful political settlement.

In mid-March 2002, Bush made what was called his first major pronouncement on the Middle East. The headlines described this as the first significant statement in years, and so on. If you read the speech, it was boilerplate, except for one sentence. That one sentence, if you take a look at it closely, said, "As progress is made toward peace, settlement activity in the occupied territories must end."[14] What does that mean? That means until the peace process reaches a point that Bush endorses, which could be indefinitely far in the future, Israel should continue to build settlements. That's also a change in policy. Up until now, officially at least, the United States has been opposed to expansion of the illegal settlement programs that make a political solution impossible. But now Bush is saying the opposite: Go on and settle. We'll keep paying for it, until we decide that somehow the peace process has reached an adequate point. This represents a significant change toward more aggression, undermining of international law, and undermining of the possibilities of peace.

You've described the level of public protest and resistance to the Iraq war as "unprecedented."[15] Never before has there been so much opposition before a war began. Where is that resistance going in the United States and internationally?

I don't know any way to predict human affairs. It will go the way people decide it will go. There are many possibilities. It should intensify. The tasks are now much greater and more serious than they were before. On the other hand, it's harder. It's just psychologically easier to organize to oppose a military attack than it is to oppose a long-standing program of imperial ambition, of which this attack is one phase, with others to come. That takes more thought, more dedication, more long-term engagement. It's the difference between deciding, I'm going out to a demonstration tomorrow and then back home, and deciding, I'm in this for the long haul. Those are choices people have to make. The same was true for people in the civil rights movement, the women's movement, and in every other movement.

What about threats to and intimidation of dissidents here inside the United States, including random roundups of immigrants and Green Card holders, and citizens, for that matter?

We definitely have to be concerned. The current government has claimed rights that go beyond any precedents, including even the right to arrest citizens, hold them in detention without access to their family or lawyers, and do so indefinitely, without charges.[16] And immigrants and other vulnerable people should certainly be cautious. On the other hand, for people like us, citizens with any privileges, though there are threats, they are so slight as compared with what people face in most of the world that it's hard to get very upset about them. I've just come back from a couple of trips to Turkey and Colombia, and compared with the threats that people face there, we're living in heaven. People in Colombia and Turkey worry about state repression, of course, but they don't let it stop them.

Do you see Europe or East Asia emerging as counterforces to U.S. power at some point?

There is no doubt that Europe and Asia are economic forces on par with North America, roughly, and have their own interests, which are not simply to follow U.S. orders. Of course, they're all tightly linked. So, for example, the corporate sectors in Europe, the United States, and most of Asia are connected in all kinds of ways and have common interests; but they also have separate interests, which is the cause of problems that go way back, especially with Europe.

The United States has always had an ambivalent attitude toward Europe. It wanted Europe to be unified, so it could serve as a more efficient market for U.S. corporations, offering great advantages of scale; but it was always concerned about the threat that Europe might move off in another direction. Many of the issues about accession of the eastern countries to the European Union (EU) are related to this. The United States is strongly in favor of this accession process, because it is hoping that these countries will be more susceptible to U.S. influence and will be able to undermine the core of Europe, which is France and Germany, big industrial countries that could move in a somewhat more independent direction.

Also in the background is a long-standing U.S. hatred of the European social system, which provides decent wages, working conditions, and benefits. The United

States doesn't want that model to exist, because it's a dangerous one. People may get funny ideas. And it's understood that the accession of eastern European countries, with economies based on low wages and repression of labor, may help undermine the social standards in western Europe. That would be a big benefit for the United States.

With the U.S. economy deteriorating and with the prospect of more layoffs on the horizon, how is the Bush administration going to maintain what some are calling a garrison state, engaged in permanent war and the occupation of numerous countries? How are they going to pull it off?

They only have to pull it off for about another six years. By that time, they hope to have institutionalized a series of highly reactionary programs within the United States. They will have left the economy in a very serious state, with huge deficits, pretty much the way they did in the 1980s. And then it will be somebody else's problem. Meanwhile, they will have undermined social programs and diminished democracy—which of course they hate—by transferring decisions out of the public arena into private hands. Internally, the legacy they leave will be painful and hard, but only for the majority of the population. The people they're concerned about are going to be making out like bandits, very much like during the Reagan years. Many of the same people are in power now, after all.

And internationally, they hope that they will have institutionalized the doctrines of imperial domination through force and preventive wars of choice. In military force and spending, the United States probably exceeds the rest of the world combined, and is now moving in extremely dangerous directions, including the militarization of space. And they assume, I suppose, that no matter what happens to the economy, U.S. military force will be so overwhelming that people will just have to do what they say.

What do you say to the peace activists in the United States who labored to prevent the invasion of Iraq and who now are feeling a sense of anger, and despair, that their government has done this?

That they should be realistic. Consider abolitionism. How long did the struggle go on before the abolitionist movement made any progress? If you give up every time you don't achieve the immediate gain you want, you're just guaranteeing that the worst is going to happen. These are long, hard struggles. And, in fact, what has happened in the last couple of months should be seen quite positively. The basis was created for expansion and development of a peace and justice movement that can go on to much harder tasks. And that's the way these things are. You can't expect an easy victory after one protest march.

COLLATERAL LANGUAGE

BOULDER, COLORADO (APRIL 5, 2003)

In recent years the Pentagon, and then the media, have adopted the term collateral damage to describe the death of civilians. Talk about the role of language in shaping and forming people's understanding of events.

It has nothing much to do with language. Language is the way we interact and communicate so, naturally, people use the means of communication to try to shape attitudes and opinions and to induce conformity and subordination. This has been true forever, but propaganda became an organized and very self-conscious industry only in the last century.

It is worth noting that this industry was created in the more democratic societies. The first coordinated propaganda ministry, the Ministry of Information, was set up in Britain during the First World War. Its "task," as they put it, was "to direct the thought of most of the world."[1] What the ministry was particularly concerned with was the mind of America, and, more specifically, the thinking of American intellectuals. Britain needed U.S. backing for the war, and the ministry's planners thought if they could convince American intellectuals of the nobility of the British war effort, then these intellectuals would succeed in driving the basically pacifist population of the United States—which wanted nothing to do with European wars, rightly—into a fit of hysteria that would get them to join the war. So its propaganda was aimed primarily at influencing American opinion. The Wilson administration reacted by setting up the first state propaganda agency here, the Committee on Public Information. This is already Orwellian, of course.

The British plan succeeded brilliantly, particularly with liberal American intellectuals. People in the John Dewey circle, for example, took pride in the fact that for the first time in history, as they saw it, a wartime fervor was created not by military leaders and politicians but by the more responsible, serious members of the community—namely, thoughtful intellectuals. In fact, the propaganda campaign succeeded within a few months in turning a relatively pacifist population into raving anti-German fanatics. The country was driven into hysteria. It reached the point that the Boston Symphony Orchestra couldn't play Bach.

Wilson had won the election in 1916 on the slogan "peace without victory," but within a couple of months he turned the United States into a country of warmongers who wanted to destroy everything German. The members of Wilson's propaganda agency included people such as Edward Bernays, who became the guru of the public relations industry, and Walter Lippmann, a leading public intellectual of the twentieth century. And they very explicitly drew on their First World War experience for their work. In their writings from the 1920s, they said that they had learned you can control "the public mind," you can control attitudes and opinions, and, in Lippmann's phrase, "manufacture consent." Bernays said that the more intelligent members of the community can direct the population through "the engineering of consent," which he considered "the very essence of the democratic process."[2]

It's interesting to look back at the 1920s, when the public relations industry really began. This was the period of Taylorism in industry, when workers were being trained to become robots and every single motion was controlled and regulated. Taylorism created highly efficient industry, with human beings being turned into automata. The Bolsheviks were very impressed with Taylorism, too, and tried to duplicate it, as did others throughout the world. But the thought-control experts soon realized that you could have not only what was called "on-job control" but also "off-job control."[3] It's a fine phrase. Off-job control means turning people into robots in every part of their lives by inducing a "philosophy of futility," focusing people on "the superficial things of life, like fashionable consumption."[4] Let the people who are supposed to run the show do so without any interference from the mass of the population, who have no business in the public arena. And from that idea grew enormous industries, ranging from advertising to universities, all very consciously committed to the belief that you must control attitudes and opinions, because the people are otherwise just too dangerous.

Actually, there are good constitutional sources for this view of the public. The founding of the country was based on the Madisonian principle that the people are just too dangerous: power has to be in the hands of what Madison called "the wealth of the nation," people who respect property and its rights and are willing to "protect the minority of the opulent against the majority," which has to be fragmented somehow.[5]

It makes perfect sense that the public relations industry developed in the more democratic societies. If you can control people by force, it's not so important to control what they think and feel. But if you lose the capacity to control people by force, it becomes necessary to control attitudes and opinions.

Today it's not so much government that exercises control, but corporations. The Reagan administration had what was called an Office of Public Diplomacy. But by that time the public was no longer willing to accept state propaganda agencies, so the Reagan Office of Public Diplomacy was declared illegal, which forced the government to use more roundabout ways to manufacture consent. Now private tyrannies—corporate systems—play the role of controlling opinions and attitudes. These corporations are not taking orders from the government but are closely linked to the government, of course. And you

don't have to speculate too much about what they're doing, because they're kind enough to tell you in their own
industry publications or in academic journals.

If you go back to 1933, for example, the liberal, progressive Wilsonian scholar Harold Lasswell, the founder
of a good bit of modern political science, wrote an article
called "Propaganda" in the *Encyclopedia of the Social Sciences.*[6] People used the term *propaganda* openly then, before the association of the word with the Nazis; now
people use various euphemisms. Lasswell's message was
that we should not succumb to "democratic dogmatisms
about men being the best judges of their own interests."
They're not. Elites are. And since people are too stupid
and ignorant to understand their best interests, we
must—because we're great humanitarians—marginalize
and control them for their own benefit. And the best way
to do this is through propaganda. There is nothing nega-
tive about propaganda, Lasswell said. It's as neutral as a
pump handle. You can use it for good or for evil. And
since we're noble, wonderful people, we'll use it for good
and to ensure that the stupid, ignorant masses remain
marginalized and separated from any decision-making
capacity. This is not the right wing that I'm talking about;
these are the liberal, progressive intellectuals.

And, in fact, you can find approximately the same

thinking in Leninist doctrines. The Nazis also picked up these ideas. If you read *Mein Kampf*, Hitler was very impressed with Anglo-American propaganda. He argued, not without reason, that propaganda won the First World War, and he vowed that next time around the Germans would be ready, too—with their own propaganda system modeled on the democracies. And since then many others have tried it. But the United States remains in the forefront because it's the most free and democratic society, so it's much more important to control attitudes and opinions here.

Can you make the leap from propaganda then, and its origins, to what's going on today with what is called Operation Iraqi Freedom?

You can read it in this morning's *New York Times*. There is an interesting article about Karl Rove, the president's manager, who teaches him what to say and do—his *minder* is what they would call him in Iraq.[7] Rove is not directly involved in the war planning, and neither is Bush. That's in the hands of other people. But his goal, he says, is "to shape perceptions of Mr. Bush as a wartime leader and to prepare for the re-election campaign that will start as soon as the war ends," so that the Republicans can push through their domestic agenda. That means tax cuts—they say for the economy, but they mean for the rich—and other programs that are designed to benefit an extremely small sector of the ultra-wealthy and privileged and that will have the effect of harming the mass of the population.

Even more significant than these short-term goals, though this is not mentioned in the *New York Times* article, is the long-term effort to destroy the institutional basis for social support systems, to eliminate programs such as Social Security that are based on the conception that people have to have some concern for one another. The idea that we should feel sympathy and solidarity, that we should care whether the disabled widow across town is able to eat, has to be driven from our minds. That's a large part of the domestic agenda, quite apart from just shifting wealth and power toward ever narrower sectors.

And the way to achieve that—since people aren't going to accept it otherwise—is to make people afraid. If people are frightened that their security is threatened, they will gravitate toward the strong leaders. They will trust the Republicans to protect them from enemies and therefore suppress their own concerns and interests. And then the Republicans will be able to drive through their domestic agenda, maybe even institutionalize it, making it very hard to reverse. So first they frighten people and then they present the president as a powerful wartime leader who is succeeding in overcoming this awesome foe—an enemy chosen precisely because it can be crushed in no time.

Iraq?

Yes, Iraq. It's laid out pretty explicitly—and it's aimed at the next presidential election. That's a significant factor in this war.

Clearly, there is a huge gap between public opinion on the Iraq war in the United States and, literally, the rest of the world. Do you attribute that to propaganda?

There is no question about it. You can trace it precisely. The campaign about Iraq took off in September 2002. This is so obvious it's even discussed in mainstream publications. The chief political analyst for United Press International, Martin Sieff, has a long article describing how it was done.[8] The drumbeat of wartime propaganda began in September, which also happened to be the opening of the midterm congressional campaign. And it had a cou-

ple of constant themes. One was that Iraq was an imminent threat to the security of the United States. We've got
to stop them now or they'll destroy us tomorrow. The second was that Iraq was behind September 11. Nobody said
that straight out; instead, they all insinuated that Iraq was
responsible. Then they said Iraq is planning new atrocities. We're really in danger, and therefore we've got to
stop them now.

Take a look at the polls. They reflected the propaganda very directly. Right after September 11 the percentage of the U.S. population that thought that Iraq was
involved was, I think, 3 percent. By now about half the
population, maybe more, believes that Iraq was responsible for September 11. Since September 2002, roughly 60
percent of the population believes that Iraq is a threat to
our security. These attitudes are closely correlated to support for war.[9] If you believe that Iraq is an imminent
threat to our security and was responsible for the September 11 atrocities and is planning new ones, then it makes
sense to say that we should go to war to stop them.

No one else in the world believes any of this. No
other country regards Iraq as a threat to its security.
Kuwait and Iran, which were both invaded by Iraq, don't
regard Iraq as a threat to their security. It's ridiculous. As
a result of the sanctions, which have killed hundreds of

thousands of people, the country has the weakest economy and the weakest military force in the region.[10] Its military expenditures are less than half those of Kuwait, which has 10 percent of Iraq's population, and well below others in the Middle East.[11] And, of course, everybody in the region knows that there is a superpower there—in effect, an offshore U.S. military base—that has hundreds of nuclear weapons and massive armed forces: Israel. In fact, after the United States takes over Iraq, it's very likely that it will increase Iraqi military forces and maybe even develop the country's weapons of mass destruction, just to counterbalance other neighboring states.

Only in the United States do people fear Iraq. This is a real achievement in propaganda. It's interesting that the United States is so susceptible to this. But, for whatever reasons, the United States happens to be a very frightened country by comparative standards. Levels of fear here on almost every issue—crime, immigration, you pick it—are just off the spectrum.

And the people in Washington know this very well. Many of them are the same people who ran the country during the Reagan years and the first Bush administration. And they're replaying the script. They pursued very regressive domestic programs that harmed the population and were very unpopular, and they succeeded in staying in political power by pushing the panic button every year. And they're doing it again now. In the United States, it's not hard to do.

You usually define things with clarity and precision, yet you say conditionally that there is something in the American character that lends itself—

In the *culture*.

What makes this culture more susceptible to propaganda?

I didn't say it's more susceptible to propaganda; it's more susceptible to fear. The United States is a frightened country. And the reasons for this—frankly, I don't understand them—probably go way back in American history.

But if the fear is there, then propaganda becomes relatively easy to implement.

Certain kinds of propaganda become much easier to implement. When my kids were in school forty years ago, during the Cold War, they were being taught literally to hide under desks to protect them from atomic bombs. Actually, there is a comment by the Mexican ambassador back at that time that ought to be famous. President Kennedy was trying to organize the hemisphere to support his terrorist attacks against Cuba, which were very severe. Generally other countries in the Western Hemisphere just have to do what they're told by the United States, or they're in bad trouble. But Mexico refused to go along with the campaign against Cuba. And the Mexican ambassador said, "If we publicly declare that Cuba is a threat to our security, forty million Mexicans will die laughing."[12]

In the United States people don't die laughing. People are afraid of everything. Take the issue of crime. The crime rate in the United States is comparable to that of other industrial societies; it's toward the high end of that scale, but not off the spectrum. Yet fear of crime here is much higher than in other countries. Drug use is about the same here as in any other country, but fear of drugs is off the scale.

But don't you think media culture contributes to that, all the television shows and movies?

Maybe, but there is also a background of fear that is exploited. It probably has to do with conquest of the continent, when you had to exterminate the native population, and slavery, when you had to control a population that was regarded as dangerous, because you never knew when the slaves might turn on you. And it may also be a reflection of the enormous security we have here. The security of the United States is unparalleled. The United States controls the hemisphere; it controls both oceans and the opposite sides of both oceans. The last time the United States was threatened was during the War of 1812. Since then, it has just conquered others. Somehow this engenders a sense that somebody is going to come after us, and the country ends up being very frightened.

Bush gave a prime-time press conference, his first in a year and a half, on Thursday, March 6, 2003. It was actually a pre-scripted press conference. He knew in advance who he was going to call on. A study of the transcript reveals a constant repetition of certain phrases—Iraq, Saddam Hussein, threat, increasing threat, deep threat, 9/11, terrorism. On the following Monday, there was a sharp spike in public opinion polls in the United States, showing a majority now believe that Iraq was connected to 9/11.

You're right about the spike, but the real change was in September 2002. That's when the poll results indicate the belief in Iraqi participation in 9/11. But that idea has to keep being reinforced, or it will just drop off. The administration's claims are so outlandish that it's very hard to expect people to stick with them unless you keep repeating them. It's the same if you're trying to sell cars. That's what you have to do. If you're trying to turn people into mindless consumers so they don't interfere with you while you're reordering the world, you have to keep at them from infancy.

How does one recognize propaganda? What are some techniques to resist it?

There are no techniques, just ordinary common sense. If you hear that Iraq is a threat to our existence, but Kuwait doesn't seem to regard it as a threat to its existence and nobody else in the world does, any sane person will begin to ask, where is the evidence? As soon as you ask this, the argument collapses. But you have to be willing to develop an attitude of critical examination toward whatever is presented to you. Of course, the whole educational system and the whole media system have the opposite goal. You're taught to be a passive, obedient follower. Unless you can break out of those habits, you're likely to be a victim of propaganda. But it's not that hard to break out.

On May 1, 1985, Reagan declared a national emergency in the United States because of the threat to the security of the United States posed by the government of Nicaragua, which was two days' drive from Harlingen, Texas, and was planning to take over the hemisphere. If you take a look at that Executive Order, which was renewed annually as a way of building up support for the U.S. war in Nicaragua, it has almost the same wording as the October 2002 congressional declaration on Iraq.[13] Just replace Nicaragua with Iraq. How much critical intelligence does it take to determine how much of a threat Nicaragua was to the existence of the United States? Again, people outside just look at this in wonder and don't understand it. Right through the 1980s, the tourist industry in Europe collapsed every few years because Americans were so frightened as a result of some spike in media coverage of terrorism that they thought, if we go to Europe there will be some Arab there who is going to try to kill us. Europeans don't know what to make of this. How can a country be so frightened of something completely nonexistent that they're afraid to travel to Europe?

That's happening again right now.

Yes, it's happening again. But in answer to the question "How do you break out of this?" just use your ordinary intelligence. There are no special techniques. Just be willing to examine what's presented to you with ordinary common sense, skeptical intelligence. Read what's presented to you the same way you would read Iraqi propaganda. Do you have to have special techniques for deciding that the minister of information in Iraq isn't to be trusted? Look at yourself the same way. If you're willing to apply to yourself the same standards you apply to others, you've won. From then on it's easy.

One of the new lexical constructions that I'd like you to comment on is "embedded journalists."

No honest journalist should be willing to describe himself or herself as "embedded." To say, "I'm an embedded journalist" is to say, "I'm a government propagandist." But journalists have accepted the term. And since anything we do is right and just, if you're a journalist embedded in an American unit, you must be objective.

The issue of embedded reporters came up dramatically in the Peter Arnett case. Peter Arnett is an experienced, respected journalist with a lot of achievements to his credit. But he's hated now because he gave an interview on Iraqi television.[14] Is anybody condemned for giving an interview on U.S. television? No, that's wonderful. From the standpoint of an independent journalist, giving an interview on U.S. television should be exactly the same as giving one on Iraqi television. In fact, it's worse; it's not a symmetrical situation. The United States is invading Iraq. It's as open an act of aggression as there has been in modern history, a major war crime. This is the crime for which the Nazis were hanged at Nuremberg, the act of aggression. Everything else was secondary. And here's a clear and open example. The pretenses for the invasion are no more convincing than Hitler's. So actually to claim that there is symmetry is already wrong, but let's put that aside. An independent journalist giving an interview over the television of the invading forces or giving an interview over the television of the invaded country shouldn't be any different, but here it's described as treachery. Arnett abandoned his journalistic integrity, and so on. What this demonstrates about U.S. journalism is astounding.

Actually, one of the best American journalists, who is therefore one of the least used, Charles Glass, a Middle East correspondent with tremendous experience, has an article in the *London Review of Books* in which he points out that the United States must be the only country in the world where someone could be called a terrorist for defending his own country from attack.[15] He's in Iraq, and he's watching this with wonder. And, in fact, anybody who is even a little bit removed from the United States and its system of indoctrination has to observe this with wonder.

The attack on Afghanistan in October 2001 generated a few other interesting terms. One was the name of the war itself, Enduring Freedom, and the other is "unlawful combatant."

After the Second World War, a relatively new framework of international law was established, including the Geneva Conventions. This framework doesn't include any such concept as "unlawful combatant" in the way it's now being used. Actually, this category predates the Second World War, when you were allowed to do just about anything during wartime. But under the Geneva Conventions, which were established to criminalize formally the atrocities of the Nazis, the situation changed. Prisoners of war are supposed to have special status. So the Bush administration, with the cooperation of the media and the courts, is going back to the period before there was any serious international framework dealing with crimes against humanity or crimes of war. Washington has claimed the right not only to carry out specific acts of aggression but to classify the people it bombs and captures as "unlawful combatants" who have no legal protections.

In fact, they have gone well beyond that. The administration has now claimed the right to round up people here, including American citizens, place them in confinement indefinitely without access to families and lawyers, and to hold them without charges until the president decides that the "war against terror," or whatever he wants to call it, is over.[16] It's astonishing. The government is claiming the right to strip people of their fundamental right of citizenship if the attorney general merely *infers*—he doesn't have to have any evidence—that the person is involved somehow in actions that might be harmful to the United States.[17] You have to go back to totalitarian states to find anything like this.

What's going on in Guantánamo, for example, is one of the worst violations of elementary principles of international humanitarian law since the Second World War, that is, since these crimes were formally criminalized in reaction to the Nazis. Even Winston Churchill, in the middle of the Second World War, condemned the use of executive power to imprison people without charge as the most odious of crimes, found only in Nazi and Communist societies. Britain was in rather desperate straits at the time, not like the United States is today. There is a bust of Churchill looking at George Bush every day. Bush might want to pay attention to his words.[18]

What do you make of Prime Minister Tony Blair of Britain being quoted on Nightline *on March 31 saying, with regard to the attack on Iraq, "This is not an invasion"?*[19]

Tony Blair is a good propaganda agent for the United States. He's articulate, his sentences hang together, apparently people like the way he looks. He's following a position that Britain has taken, self-consciously, since the end of the Second World War. During the Second World War, Britain recognized—we have plenty of internal documents about it—the obvious: Britain had been the world-dominant power, but the United States was going to become the dominant power after the war. Britain had to make a choice. Was it going to be just another country, or was it going to be what they called a "junior partner" of the United States? It accepted the role of junior partner. And that's what it's been since then. Britain has been kicked in the face over and over again in the most disgraceful way, and Blair sits there quietly and says, "We will be the junior partner." We will bring to the "coalition" our experience of centuries of brutalizing and murdering foreign people. We're good at that. We've got centuries of experience in what Lloyd George called "bomb[ing] niggers."[20] We'll be the junior partner, and maybe in return we'll get some privileges. And that's the British role. It's disgraceful.

At the talks you give to American audiences, you often are asked the question, "What should I do?"

Only by American audiences. I'm never asked this in the third world. When you go to Turkey or Colombia or Brazil, they don't ask you, "What should I do?" They tell you what they're doing. When I went to Porto Alegre, Brazil, for the World Social Forum, I met with some landless *campesinos,* and they didn't ask me what they should do; they told me what they were doing. These are poor, oppressed people, living under horrendous conditions, and they would never dream of asking you what they should do. It's only in highly privileged cultures like ours that people ask this question. We have every option open to us, and have none of the problems that are faced by intellectuals in Turkey or *campesinos* in Brazil. We can do anything. But people here are trained to believe that there are easy answers, and it doesn't work that way. If you want to do something, you have to be dedicated and committed to it day after day. Educational programs, organizing, activism. That's the way things change. You want a magic key, so you can go back to watching television tomorrow? It doesn't exist.

You were an active and early dissident in the 1960s, opposing U.S. intervention in Indochina. How has dissent evolved in the United States since that time?

It's kind of interesting. There was an article in the *New York Times* this morning describing how it's the professors who are antiwar activists today, not the students.[21] It's not like it used to be, when the students were the antiwar activists. It's true that by 1970 students were active antiwar protesters. But that only happened after eight years of the U.S. war against South Vietnam, which by then had been extended to all of Indochina and had practically wiped the place out.

In 1962, it was announced that U.S. planes were bombing South Vietnam—there was no protest. The United States used chemical warfare to destroy food crops and drive millions of people into "strategic hamlets," essentially concentration camps. All of this was public, but there was no protest; it was impossible to get anybody to talk about it. Even in a liberal city like Boston, you couldn't have public meetings against the war because they would be broken up by students, with the support of the media. You would have to have hundreds of state police around to allow speakers like me to escape unscathed. The protests came only after years and years of war. By then, hundreds of thousands of people had been killed and much of Vietnam had been destroyed.

But all of that is erased from history, because it tells too much of the truth, which is that it took years and years of hard work by plenty of people, mostly young, to build a protest movement. But the *New York Times* reporter can't understand that. I'm sure she's being and saying exactly what she was taught, that there was a huge antiwar movement and now it's gone. The actual history can't be acknowledged. You aren't supposed to learn that dedicated, committed effort can bring about significant changes of consciousness and understanding. That's a very dangerous idea, and therefore it's been wiped out of history.

REGIME CHANGE

CAMBRIDGE, MASSACHUSETTS (SEPTEMBER 11, 2003)

Regime change is a new term in the lexicon, but the United States is an old hand at regime change. This year there are several anniversaries. Today is the thirtieth anniversary of the U.S.-backed coup in Chile. October 25, 2003, will mark the twentieth anniversary of the U.S. invasion of Grenada. But I'm particularly thinking of the regime change in Iran fifty years ago, in August 1953, which overturned the conservative parliamentary democracy led by Mohammed Mossadegh and restored the shah, who ruled for the next twenty-five years.

The issue in Iran was that a conservative nationalist parliamentary government was attempting to take back its own oil resources. These had been under the control of a British company—originally Anglo-Persian, later named Anglo-Iranian—which had entered into contracts with the rulers of Iran that were pure extortion and robbery. The contracts gave the Iranians nothing, and the British were laughing all the way to the bank.

Mossadegh was a long-standing critic of this subordination to imperial policy. Popular outbursts compelled the shah to appoint him as prime minister, and he moved to nationalize the industry, which made perfect sense. The British went completely berserk. They refused to make any compromises like the ones American oil companies had just agreed to in Saudi Arabia. They wanted to continue robbing the Iranians blind. And that led to a tremendous popular uprising in support of nationalization.

Iran had a long democratic tradition, including a *majlis*, a parliament. And the shah couldn't suppress it. Finally a joint British-American coup succeeded in overthrowing Mossadegh and restoring the shah to power, ushering in twenty-five years of terror, atrocities, and violence, which led finally to the revolution in 1979 and the expulsion of the shah.

Incidentally, one outcome of the 1953 coup was that the United States took over about 40 percent of Britain's share in Iranian oil. That wasn't the goal of the effort—it just happened in the normal course of events—but it was part of the general displacement of British power by U.S. power in that region, and in fact throughout the world. The *New York Times* ran an editorial praising the coup, in which it said, "Underdeveloped countries with rich resources now have an object lesson in the heavy cost that must be paid by one of their number which goes berserk with fanatical nationalism."[1] Other Mossadeghs elsewhere in the world should be careful before trying to do something like gaining control of their own resources—which, of course, are ours, not theirs.

But your point is quite correct. Regime change is normal policy. If you go back to the Kennedy and Johnson administrations, there was a period of real frenzy about regime change in Cuba. Internally, the reason given by U.S. intelligence for regime change was that the very existence of the Castro regime "represents a successful defiance of the United States, a negation of our whole hemispheric policy of almost a century and a half," meaning the Monroe Doctrine.[2] So we have to overthrow Cuba by a campaign of large-scale terror and economic warfare. This terrorist campaign almost led the world to a terminal nuclear war. It was very close.

Right after the First World War, the British replaced the Turks as the rulers of Iraq. They occupied the country, and faced, as one account says, "anti-imperialist agitation . . . from the start." A revolt "became widespread." The British felt it wise to put up an "Arab façade," as Lord Curzon, the foreign secretary, called it, "ruled and administered under British guidance and controlled by a native Mohammedan, and, as far as possible, by an Arab staff."³ Fast-forward to Iraq today, with a twenty-five-person ruling council appointed by the U.S. viceroy, L. Paul Bremer III.

Lord Curzon was very honest in those days. Iraq would be an Arab façade. Britain's rule should be "veiled" behind such "constitutional fictions as a protectorate, a sphere of influence, a buffer State, and so on."⁴ And that's the way Britain ran the whole region—in fact, the whole empire. The idea is to have independent states, but with weak governments that must rely on the imperial power for their survival. They can rip off the population if they like. That's fine. But they have to provide a façade behind which the real power can rule. That's standard imperialism.

You can find plenty of examples. The current occupation of Iraq is one. There was a wonderful organizational chart published in the *New York Times* last May, just after Bremer was appointed.[5] Unfortunately it's not in the archived electronic edition, so you have to look back at the hard copy or look it up on microfilm, but it was a standard organizational chart with something like seventeen boxes. The person at the top is Paul Bremer, who answers to the Pentagon. Below Bremer, you have lines to the various generals and diplomats, all either U.S. or British, with the responsibilities of their office listed in boldface. Then you get down to the bottom and there's a seventeenth box, half the size of the others, with no boldface and no indication of responsibility. And this box says, "Iraqi Advisers." That expresses the thinking— that's the façade. Lord Curzon would have considered it quite normal.

I should say, though, to my amazement, the occupation is not succeeding. It takes real talent to fail in this. For one thing, military occupations almost always work. At the extreme end of the spectrum of brutality, the Nazis in occupied Europe had very little trouble running the countries under their control. Every country had a façade of collaborators who kept order and kept the population down. If the Nazis hadn't been crushed by overwhelming outside force, they wouldn't have had any trouble continuing to run occupied Europe. The Russians, who were also extremely brutal, had very little problem running eastern Europe through façades.

Furthermore, Iraq is an unusually easy case. Here is a country that has been decimated by a decade of murderous sanctions that killed hundreds of thousands of people and left the whole place in tatters, devastated by wars, and run by a brutal tyrant. The idea that you can't get a military occupation to run under these circumstances, and with no support from outside for the resistance, is almost inconceivable. I imagine if we got a couple of people on this floor together here at the Massachusetts Institute of Technology, we could probably figure out how to get the electricity working, but the U.S. occupation hasn't. The occupation of Iraq has been an astonishing failure. The administration's original planning, as illustrated in that organization chart, amazingly looks like it isn't going to work. Which is why you now hear all this backtracking about trying to get the United Nations to come in and pick up some of the costs. It's a big surprise to me. I thought this would be a walkover.

Jawaharlal Nehru, one of the leaders of the opposition to British rule in India, observed that the ideology of British rule in India "was that of the herrenvolk and the Master Race," an idea that is "inherent in imperialism." These racist ideas were "proclaimed in unambiguous language by those in authority" and "Indians as individuals were subjected to insult, humiliation, and contemptuous treatment."[6] Is racism "inherent" in imperialism?

It's worth remembering that Nehru was an Anglophile. But even for Nehru—who was from the elite Indian upper classes and quite British in manner and style—the humiliation and degradation were hard to bear. Nehru is right. Racism is inherent in imperial rule—it's almost invariable. And I think you can understand the psychology. When you have your boot on somebody's neck, you can't just say, "I'm doing this because I'm a brute." You have to say, "I'm doing it because they deserve it. It's for their good. That's why I've got to do it." They're "naughty children," who have to be disciplined.[7] Filipinos were described in the same way. And it's exactly what's been going on in the Palestinian Occupied Territories for years. One of the worst aspects of the Israeli occupation has been the humiliation and degradation of Palestinians at every moment. That's inherent in the relation of domination.

What about the drive for resources?

That's a very consistent factor in domination, but it's not always the only factor. For example, the British didn't want to control Palestine for its resources but for its geostrategic position. Many factors enter into the ambition for domination and control, but the drive for resources is a very common one. Consider the U.S. takeover of Texas and around half of Mexico about 150 years ago. That's usually not called a resource war, but it was. Look back at the Jacksonian Democrats, such as James K. Polk and other people at the time. They were trying to do exactly what Saddam Hussein was accused of trying to do in 1990 when he invaded Kuwait—to gain a monopoly over the world's major resource, which in those days was cotton—except they were open about it. Cotton fueled the industrial revolution in the same way oil now fuels the industrial world. One of the goals in taking over these territories at the time, particularly Texas, was to ensure that the United States could gain a monopoly of cotton and bring the British to their knees—because we would control the resource on which they survived. Britain was the world's leading industrial power and the United States was then a minor industrial power. And remember, Britain was the great enemy at the time, a powerful force that was preventing the United States from expanding north to Canada and south to Cuba. So it was a resource war, in a deep sense, though there were other factors at play. It's not unusual to find that. The Israeli takeover of the West Bank, for example, is partially for water resources, which Israel needs, but the reasons go way beyond that.

Why did the United States attack Iraq, which posed no threat, rather than North Korea, which has a far more developed military and nuclear program?

Iraq was completely defenseless, whereas North Korea had a deterrent. The deterrent is not nuclear weapons. The deterrent is the massed artillery at the Demilitarized Zone, aimed at Seoul, the capital of South Korea, and at maybe tens of thousands of American troops at the border. Unless the Pentagon can figure out some way of taking out that artillery with precision-guided weapons, North Korea has a deterrent. Iraq had nothing. The Bush administration knew perfectly well that Iraq was defenseless. They probably knew where every pocketknife was in every square inch of Iraq by the time of the invasion.

Still, Korea is a major concern for the United States, in large part because of its position within Northeast

Asia. The Northeast Asian region is the most dynamic economic region in the world. It includes two major industrial societies, Japan and South Korea, and China is increasingly becoming an industrial society. It has enormous resources. Siberia has all kinds of resources, including oil. Together, the countries in Northeast Asia have close to a third of the world's gross domestic product, way more than the United States, and about half of global foreign exchange. The region has enormous financial resources. And it's growing very fast, much faster than any other region including the United States.[8] Its trade is increasing internally and it's connecting to the Southeast Asian countries, sometimes called ASEAN Plus Three: the countries in the Association of South East Asian Nations plus China, Japan, and South Korea. Some of the pipelines being built from the resource centers to the industrial centers would naturally go to South Korea, which means right through North Korea. If the Trans-Siberian railway is extended, as is surely planned, it will probably follow the same route through North Korea to South Korea. So North Korea is in a fairly strategic position with regard to this area.

The United States is not particularly happy about Northeast Asian economic integration, in much the same way it has always been ambivalent about European integration. It has always been a concern. Quite a lot of policy planning, from the Second World War to the present, reflects the concern that Europe might take an independent course; it might be what used to be called a "third force." That's a lot of the purpose of the North Atlantic Treaty Organization, in fact. The same issues are arising for Northeast Asia today. So the world now has three major economic centers: North America, Northeast Asia, and Europe. In one dimension, the military dimension, the United States is in a class by itself—but not in the others.

Zbigniew Brzezinski, Jimmy Carter's national security adviser, contends that "the three grand imperatives of [U.S.] imperial geostrategy are to prevent collusion and maintain security dependence among the vassals, to keep tributaries pliant and protected, and to keep the barbarians from coming together."[9]

That's pretty frank—and it's basically correct. Lord Curzon would have been pleased. In international relations theory, this is called "realism." You prevent other powers from grouping together to oppose the hegemonic power. Part of the reason that conservative international relations specialists like Samuel Huntington and Robert Jervis were highly critical of U.S. policy was the observation that U.S. policies were creating a situation in which much of the world regarded the United States as a "rogue state," a threat to their existence, and would form coalitions against U.S. hegemony. And this was in the Clinton years, before the Bush administration's National Security Strategy.

In a 1919 essay called "The Sociology of Imperialisms," the Austrian economist Joseph Schumpeter wrote:

> There was no corner of the world where some interest was not alleged to be in danger or under actual attack. If the interests were not Roman, they were of Rome's allies; and if Rome had no allies, then allies would be invented. When it was utterly impossible to contrive such an interest—why, then it was the national honor that had been insulted. The fight was always invested with an aura of legality. Rome was always being attacked by evil-minded neighbors, always fighting for a breathing space. The whole world was pervaded by a host of enemies, and it was manifestly Rome's duty to guard against their indubitably aggressive designs.[10]

Monthly Review used that quote in a fairly recent issue in an editorial referring to Bush's National Security Strategy, precisely because it is so apposite.[11] You just change the words from Rome to Washington. One of the standard arguments for going to war these days is to "maintain credibility." In some cases credibility is at stake—not resources. Take, say, the bombing of Serbia in 1999, again under Clinton. What was the point of that? The standard line is that the United States intervened to prevent ethnic cleansing, but to hold to that you have to invert the chronology. Uncontroversially, the worst ethnic cleansing followed the bombing and, furthermore, was the anticipated consequence of it. So that can't have been the reason. What was the reason? If you look carefully, Clinton and Blair said at the time—as it's now retrospectively conceded—that the point of the bombing was to maintain credibility. To make clear who's the boss. Serbia was defying the orders of the boss, and you can't let anyone do that. Like Iraq, Serbia was defenseless, so there was no risk. In fact, you can proclaim how you intervened only for humanitarian reasons.

This logic should be familiar to anyone who watches television programs about the mafia. The don has to make sure that people understand he's the boss. You don't cross him. He sends out goons to beat somebody to a pulp—not because he want his resources, but because the guy's standing up to him. It was Castro's successful defiance of the United States that made it necessary to carry out terrorist actions aimed at regime change. You don't defy the master, and everyone has to understand that. If the rumor spreads that you can defy the master and get away with it, he's in trouble.

The historian William Appleman Williams in his book Empire as a Way of Life *writes, "Very simply, Americans of the twentieth century liked empire for the very same reasons their ancestors had favored it in the eighteenth and nineteenth century. It provided them with renewable opportunities, wealth, and other benefits and satisfactions, including a psychological sense of well-being and power."[12] What do you think of his analysis?*

Williams's comments are partly correct but remember that the United States was not an empire in the European style. The English colonists who came to the United States didn't create a façade of the native population behind which they would rule, like the British in India. They largely wiped out the native population—*exterminated* is the word the Founding Fathers used. And this was considered a perfectly fine thing to do. The United States was first a kind of settler state rather than an imperial state.

Subsequent territorial expansions, at least up to

World War II, followed pretty much the same pattern. Think of Mexico, large parts of which we took over in the 1840s, or Hawaii, which was stolen by force and guile in 1898. In both cases the native population was pretty much replaced, they weren't colonized. Again, not totally replaced. The indigenous people are still there, but they've essentially been taken over.

Also, if you look at the traditional empires, say, the British empire, it's not so clear that the population of Britain gained from it. It's a very difficult topic to study, but there have been a couple of attempts. And for what it's worth, the general conclusion is that the costs and the benefits pretty much balanced out. Empires are costly. Running Iraq is not cheap. Somebody's paying. Somebody's paying the corporations that destroyed Iraq and the corporations that are rebuilding it. In both cases, they're getting paid by the U.S. taxpayer. Those are gifts from U.S. taxpayers to U.S. corporations.

I don't understand. How did corporations like Halliburton and Bechtel contribute to the destruction of Iraq?

Who pays Halliburton and Bechtel? The U.S. taxpayer. The same taxpayers fund the military-corporate system of weapons manufacturers and technology companies that bombed Iraq. So first you destroy Iraq, then you re-build it. It's a transfer of wealth from the general popula-tion to narrow sectors of the population. Even if you look at the famous Marshall Plan, that's pretty much what it was. It's talked about now as an act of unimaginable benevolence. But whose benevolence? The benevolence of the U.S. taxpayer. Of the $13 billion of Marshall Plan aid, about $2 billion went right to the U.S. oil compa-nies.[13] That was part of the effort to shift Europe from a coal-based to an oil-based economy, and to make Europe-an countries more dependent on the United States. Eu-rope had plenty of coal. It didn't have oil. So there's two of the thirteen billion. If you look at the rest of the aid, very little of the money left the United States. It just moved from one pocket to another. The Marshall Plan aid to France just about covered the costs of the French effort to reconquer Indochina. So the U.S. taxpayer wasn't re-building France. They were paying the French to buy American weapons to crush the Indo-Chinese. And they were paying Holland to crush the independence move-ment in Indonesia.

Returning to the British empire, the costs to the British people may have been about on a par with the benefits that the British people received from it, but for the guys who were running the East India Company the empire led to fantastic wealth. For the British troops who were dying out in the wilderness somewhere, the costs were serious. To a large extent, that's the way empires work. Internal class war is a significant element of empire.

It's relatively easy to measure the cost in lives, the number of soldiers killed, and how much money is spent. How does one measure or even talk about moral degradation?

You can't measure that, but it's very real and very significant. And that's part of the reason that an imperial system, or any system of domination, even a patriarchal family, always has a cover of benevolence. We're back to racism again. Why do you have to present yourself as somehow doing it for the benefit of the people you're crushing? Well, otherwise you have to face moral degradation. If we're honest about it, human relations are often like that. And in imperial systems, almost always. It's hard to find an imperial system in which the intellectual class didn't laud its own benevolence. When Hitler was dismembering Czechoslovakia, it was accompanied by wonderful rhetoric about bringing peace to the ethnic groups who were in conflict, making sure they could all live happily together under benign German supervision.

You really have to labor to find an exception to that. And of course it's true in the United States.

Traditionally if you used the word imperialism *and attached "American" in front of it, you were dismissed as a member of some far left fringe. That has undergone a bit of a transformation in the last few years. For example, Michael Ignatieff, director of the Carr Center at the Kennedy School of Government at Harvard, wrote in a* New York Times Magazine *cover story that "America's empire is not like the empires of times past, built on colonies, conquest and the white man's burden.... The twenty-first-century imperium is a new invention in the annals of political science, an empire lite, a global hegemony whose grace notes are free markets, human rights and democracy, enforced by the most awesome military power the world has ever known."[14]*

Of course, the apologists for every imperial power have said the same thing. So you can go back to John Stuart Mill, one of the most outstanding Western intellectuals. He defended the British empire in very much those words. Mill wrote the classic essay on humanitarian intervention.[15] Everyone studies it in law schools. He argued that Britain is unique in the world. It's unlike any country in history. Other countries have crass motives and seek gain and so on, but the British act only for the benefit of others. In fact, he said, our motives are so pure that Europeans can't understand us. They heap "obloquy" upon us, and seek to discover crass motives behind our benevolent actions. But everything we do is for the benefit of the natives, the barbarians. We want to bring them free markets and honest rule and freedom and all kinds of wonderful things. I'm surprised Ignatieff is not aware that he's just repeating very familiar rhetoric.

The timing of Mill's comments is interesting. He wrote this essay around 1859, right after an event that in British terminology is called the "Indian Mutiny"— meaning the barbarians dared to raise their heads. The Indians launched a rebellion against British rule, and the British put it down with extreme violence and brutality. Mill certainly knew about this. It was all over the press. Old-fashioned conservatives, like Richard Cobden, condemned the British repression of the mutiny harshly, much like Senator Robert Byrd condemns the invasion of the Iraq today. The real conservatives are different from the ones who call themselves conservatives. But Mill, right in the midst of the suppression of the rebellion, wrote about Britain as an angelic power.

And people believe the rationales. If you examine the internal record, political leaders often talk to each other the same way they talk in public. For example, many doc-uments from the Soviet archives are coming out now; they're basically being sold to the highest bidder like everything else in Russia. If you look at the discussions from the 1940s, after the Second World War, you see An-drey Gromyko and other Soviet leaders discussing how they have to intervene to protect democracy from the forces of fascism, which are everywhere. I'm sure Gromyko believed what he was saying as much as Ignatieff believes what he is saying.

In another New York Times Magazine *article, Ignatieff wrote, "New rules for intervention, proposed by the United States and abided by it, would end the canard that the United States, not its enemies, is the rogue state." You have a book called* Rogue States.[16] *Is the United States a rogue state?*

Actually, I borrowed the phrase from Samuel Hunting-ton. In *Foreign Affairs*, the main establishment journal, he wrote that much of the world regards the United States as a "rogue superpower," and "the single greatest external threat to their societies."[17] Huntington was criticizing Clinton administration policies that were leading other countries to build up coalitions against the United States. If we define "rogue state" in terms of any principle, such as violation of international law, or aggression, or atrocities, or human rights violations, the United States certainly qualifies, as you would expect of the most powerful state in the world. Just as Britain did. Just as France did. And intellectuals in every one of these empires wrote the same kind of garbage that you have quoted from Ignatieff. So France was carrying out a "civilizing mission" when the minister of war was saying they were going to have to exterminate the natives in Algeria. Even the Nazis used this rhetoric. You go to the absolute depths of depravity, and you'll find the same sentiments expressed. When the Japanese fascists were conquering China and carrying out huge atrocities like the Nanking Massacre, the rhetoric behind it brings tears to your eyes. They were creating an "earthly paradise" in which the peoples of Asia would work together. Japan would protect them from the Communist "bandits" and would sacrifice itself for their benefit so they would all have peace and prosperity.[18] Again, I'm a little surprised that some editor at the *New York Times* or a distinguished professor at Harvard doesn't see that it's a little odd to just be repeating what's been said over and over again by the worst monsters. Why is it different now?

Notice, by the way, that one of the great benefits of being a respectable intellectual is that you never need any evidence for anything you say. Go through those articles, and try to find some evidence to support the conclusions. In order to make it to the peak of respectability, you have to understand that it's faintly absurd even to ask for evidence for praise of those with power. It's just automatic. Of course they're magnificent. Maybe they made some mistakes in the past, but now they're magnificent. And to look for evidence of that is like looking for evidence for the truths of arithmetic. It's as if you wrote that two plus two is four, and then somebody said, "Where's your evidence?" So there never is any.

The Italian socialist Antonio Gramsci wrote, "A main obstacle to change is the reproduction by the dominated forces of elements of the hegemonic ideology. It is an important and urgent task to develop alternative interpretations of reality."[19] How does someone develop "alternative interpretations of reality"?

I deeply respect Gramsci, but I think it's possible to paraphrase that comment—namely, just tell the truth. Instead of repeating ideological fanaticism, dismantle it, try to find out the truth, and tell the truth. It's something any one of us can do. Remember, intellectuals internalize the conception that they have to make things seem complicated. Otherwise what are they around for? It's worth asking yourself what's really so complicated? Gramsci is a very admirable person, but take that statement and try to translate it into simple English. How complicated is it to understand the truth or to know how to act?

WARS OF AGGRESSION

CAMBRIDGE, MASSACHUSETTS (FEBRUARY 12, 2004)

In a new documentary The Fog of War, *Robert McNamara makes a rather interesting admission. He quotes General Curtis LeMay, with whom he served in the period of the firebombing of Japanese cities in World War II, as saying, "If we'd lost the war, we'd all have been prosecuted as war criminals." Then McNamara says, "I think he's right. . . . But what makes it immoral if you lose and not immoral if you win?"*[1]

I haven't seen the film, but I've been told that in it McNamara identifies his own role during the Second World War for the first time. The biographical material typically describes him as kind of a statistician who was working somewhere in the background, but it turns out that he was actually in a planning role, figuring out how to maximize Japanese civilian deaths at minimal cost. Apparently, Tokyo was selected as a target because it was very densely populated and made mostly of wood, so you could start a firestorm that would kill some one hundred thousand people with no difficulty. Remember that Japan had no air defenses at this point. I understand that McNamara takes responsibility—I can't say credit, exactly— for having made this decision.

His comment about war criminals is not only true in this example, but in general. Telford Taylor, who was chief prosecutor at the Nuremberg war crimes tribunal, pointed out that the tribunal was prosecuting post facto crimes, that is, crimes that were not on the books at the time they occurred.[2] The tribunal had to decide what would be considered a war crime, and they made the operational definition of a war crime anything the enemy did that the Allies didn't do. This was explicit—and it explains why, for example, the devastating Allied bombings of Tokyo, Dresden, and other urban civilian centers were not considered war crimes. The U.S. and British air forces did much more bombing of urban civilian centers than did the Germans. They aimed mainly at working-class and poor civilian areas. But since the Allies did it much more than the Axis, bombing urban centers was removed from the category of war crimes. That same principle showed up in individual testimonies as well. A German admiral—Karl Doenitz, the submarine commander—brought as a defense witness an American submarine commander, Nimitz, who testified that Americans had done the things that Doenitz was charged with. He was exonerated.

The Nuremberg tribunal was at least semi-respectable. The Tokyo tribunal was simply a farce. And some of the other trials of the Japanese were just unbelievable, like the trial of General Tomoyuki Yamashita, who was charged and hanged for crimes committed by Japanese soldiers in the Philippines. The soldiers were technically under his command, but at the end of the war they were cut off, and he had no communication with them. They did commit terrible atrocities. And he was hanged for it.[3] Just imagine if that example were generalized to commanders whose soldiers, on their own, without any direct communication, committed crimes. The whole military command of every functioning army in the world would be hanged, as would the civilian leadership. It's not the generals, it's the civilians who usually authorize and organize the worst war crimes. So McNamara's observation is accurate, familiar, and an understatement.

Incidentally, McNamara's point applies to war crimes trials that are happening today. You recall the reaction when for about thirty seconds it looked as though the special tribunal for Yugoslavia might investigate NATO crimes. Canadian and British lawyers urged the tribunal to look into NATO war crimes—which of course took place—and for a brief moment it looked as if it might. But the United States quickly warned the tribunal that it had better not pursue any U.S. or allied crimes. Crimes are something others do, not something we do.

The same logic can be found in the Bush doctrine. One component of the doctrine is that the United States has the right to carry out offensive military actions against countries we regard as a security threat because they have weapons of mass destruction. That's the first part of the doctrine. Many establishment figures criticized it not so much because they disagreed with it but because they thought the brazenness of its declaration and implementation was ultimately a threat to the United States. *Foreign Affairs* immediately published a critical article on what it called the "new imperial grand strategy."[4] Even Madeleine Albright, the Clinton secretary of state, pointed out, quite accurately, that while every president has had such a doctrine you don't advertise it. "Anticipatory self-defense," she wrote in *Foreign Affairs*, is "a tool every president has quietly held in reserve."[5] You keep it in your back pocket, and you use it when you want to. The most interesting comment, perhaps, was Henry Kissinger's, responding to a major address by President Bush at West Point in which he had presented an outline of the National Security Strategy. Kissinger said this "revolutionary" doctrine in international affairs would tear to shreds not only the UN Charter and international law but the whole seventeenth-century Westphalian system of international order. Kissinger approved of the doctrine, though he added one proviso: we have to understand that this can't be "a universal principle available to every nation."[6] The doctrine is for us, not for anyone else. We will use force whenever we like against anyone we regard as a potential threat, and maybe we will delegate that right to client states, but it's not for others.

Let's turn to the second part of the Bush doctrine: "Those who harbor terrorists are as guilty as the terrorists themselves."[7] Just as we have the right to attack and destroy terrorists, we have the right to attack and destroy states that harbor terrorists. Okay, which states harbor terrorists? Let's put aside those states that are harboring heads of state; if we include them, the discussion reduces to absurdity in no time. Let's restrict ourselves to those groups and individuals officially regarded as terrorists or subnational terrorists such as Al Qaeda or Hamas. Which states harbor them? Right now there is an extremely important case coming to an appeals court in Miami that bears on this question very directly, the case of the Cuban Five. I haven't seen much coverage of it. Just to give a little background, the United States launched a terrorist war against Cuba in 1959, which picked up rapidly under Kennedy, with Operation Mongoose, and actually came close to triggering a nuclear war. The peak of the atrocities was probably in the late 1970s. By that time, though, the United States was dissociating itself from the terrorist war and, as far as we know, was not carrying out terrorist actions directly. Instead, the United States was harboring terrorists who were carrying out attacks on Cuba—quite serious ones—in violation of U.S. and international law. The terrorist acts, incidentally, continued at least into the late 1990s. We don't have to debate about whether the people involved are terrorists are not. The FBI and the Justice Department describe them as dangerous terrorists, so let's take their word for it. There's Orlando Bosch, for example, whom the FBI accuses of numerous serious terrorist acts, some of them on U.S. soil, and whom the Justice

Department described as a threat to the security of the United States who should be deported. Bosch's activities include participation in the destruction of a Cubana airliner, in which seventy-three people were killed, in 1976. George Bush I, at the request of his son Jeb, the Florida governor, gave Bosch a presidential pardon.[8] So he's sitting happily in Miami, and we're harboring a person whom the Justice Department regards as a dangerous terrorist, a threat to the security of the United States.

When it became clear that the United States was doing nothing to stop terrorists harbored here from carrying out attacks, Cuba decided to infiltrate the terrorist organizations in Florida with agents of its own to collect information. Cuba then invited FBI agents to come to Havana, which they did. In 1998, Cuba provided high-level FBI officials with thousands of pages of documents and videotapes about the planning of terrorist actions in Florida. And the FBI responded, namely, by arresting the infiltrators. That's the case of the Cuban Five: the infiltrators who gave the FBI the information about terrorists in the United States were arrested. They were brought to court in Miami, and the judge refused a change of venue, which is ridiculous. The prosecutor conceded that there was basically no case against the Cubans, but they were convicted anyway. The case is being appealed, but three of them have life sentences, the others long sentences, and their families have been denied the right to visit them.[9] This is a perfect example of a state harboring terrorists—and should be a major scandal.

This is not the only example. The Venezuelan government is now seeking extradition of two military officers who were accused of participation in bombing attacks in Caracas, fled the country, and now are pleading for political asylum here.[10] These officers participated in a military coup in 2002, which succeeded for a couple of days in ousting the Chávez government. The U.S. government openly supported the coup and, according to quite good journalists in the British press, was involved in instigating it.[11] If some military officers in the United States had taken over the White House and run the government, they would have been executed. But the very reactionary Venezuelan courts, which are still tied to the old regime, refused the government's efforts to try the officers. The "totalitarian" Chávez regime agreed to the court ruling and didn't try them. So they were set free. Now they are seeking asylum in the United States, and I assume they will receive it.

Or take Emmanuel Constant. He is responsible for killing maybe four or five thousand Haitians. He is living happily in Queens, New York, because the United States refuses even to respond to requests for extradition.[12]

So who is harboring terrorists? If states that harbor terrorists are terrorist states, according to the Bush doctrine, what do we conclude? We conclude exactly what Kissinger was kind enough to say: such doctrines are unilateral. They are not intended as norms of international law; they are doctrines that grant the United States the right to use force and violence and to harbor terrorists, but not anyone else. For the powerful, crimes are those that others commit.

Robert Jackson, the chief U.S. prosecutor at Nuremberg, said in his opening speech that "to start or wage an aggressive war has the moral qualities of the worst of crimes."[13] His British counterpart, Hartley Shawcross, said the Germans had committed a "crime against peace . . . waging wars of aggression and in violation of Treaties."[14] Under the United Nations Charter, the planning and waging of aggressive war is regarded as a major war crime.[15] Given the attack on Iraq, a country that was not threatening the United States, why hasn't there been any discussion about the U.S. government waging an illegal war of aggression? And why aren't people talking about impeaching President Bush?

They are. Various lawyers' groups in the United States—
but mostly in England, Canada, and elsewhere—are
seeking to put U.S. officials on trial for the crime of aggression. We should point out, however, that while the invasion of Iraq was plainly an act of aggression, it wasn't
unprecedented. What was the 1962 invasion of South
Vietnam, for example, when Kennedy sent the air force to
attack South Vietnam and began a campaign of chemical
warfare, with devastating consequences, driving the population into concentration camps? That was aggression.
You could say it was aggression against a state that was
not a member of the United Nations, if that matters, but it
was certainly aggression. Or what was the Indonesian invasion of East Timor? Obviously aggression. Or the Israeli invasion of Lebanon, which ended up killing twenty
thousand people?[16] Both of these were carried out thanks
to decisive U.S. diplomatic, military, and economic support. In the case of East Timor, Britain was also involved.
And we can go on.

The 1989 invasion of Panama, for example, What was
that? An invasion aimed at kidnapping a thug, not a thug
of Saddam Hussein's ranking but a serious one, Manuel
Noriega. In the course of the invasion, the U.S. military
killed, according to Panamanian sources, three thousand

civilians.[17] We can't confirm the number because we don't investigate our own crimes. Nobody knows for certain, but the U.S. invasion of Panama certainly killed plenty of people—on the scale of the Iraqi invasion of Kuwait, with roughly the same number of casualties. The United States vetoed Security Council resolutions and General Assembly resolutions condemning the invasion.[18] Noriega was seized from the Vatican embassy and brought back to Florida—all hopelessly illegal—and then, in a ridiculous trial, he was convicted of crimes that he had indeed committed, almost all of them when he was on the CIA payroll.[19] If Saddam Hussein ever comes to trial, it will be the same: he will be convicted of crimes that the U.S. supported, but that fundamental detail won't be mentioned.

How does the international law community deal with this? International law professionals have a complicated task. There is a fringe who tell the truth and point out the violations of international law. But most have to construct complex arguments to justify crimes of aggression. Their job, basically, is to serve as defense counsels for state power. Their justifications are interesting. The more honest people, like Michael Glennon of the Fletcher School of Law and Diplomacy, simply say that international law and the UN Charter are a lot of "hot air," and they should be eliminated because they restrict the ability of the United States to use force.[20]

Glennon's position—which is shared by many other defenders of U.S. aggression, such as Yale University law professor Ruth Wedgwood—is that U.S. actions like the illegal bombing of Serbia have changed the nature of law, because law is a living doctrine, a living system of principles, which is continually modified by international practice. Was it modified by Saddam Hussein's invasion of Kuwait? No. Was it modified by Vietnam's invasion of Cambodia, one of the few actions in modern history that might properly be called a humanitarian intervention? Or India's invasion of East Pakistan, which put an end to huge atrocities? No. In fact, these interventions were all bitterly condemned. None of them created new norms of international law. And that's because we are the ones who change the law, not anybody else.

A recent issue of the *American Journal of International Law* has a complex, thoughtful article by Carsten Stahn called "Enforcement of the Collective Will After Iraq." Stahn quotes Jürgen Habermas and all sorts of other big thinkers. His argument comes down to this: When the United States invaded Iraq, it actually was abiding by the UN Charter, if one interprets it prop-

erly. We have to recognize that there are two interpretations of the charter. There's a literal interpretation, that
the use of force in international affairs is criminal except under circumstances that didn't apply in the case
of Iraq, which is trivial and uninteresting. Then there is
the "communitarian" interpretation of the charter, that
an act is legitimate if it carries out the will of the community of nations. Since the Security Council doesn't
have the military force to carry out the will of the community of nations, it implicitly delegates this role to
states that do have the force, meaning the United
States. And therefore, under the communitarian interpretation of the charter, the United States, by invading
Iraq, was fulfilling the will of the international community. It's irrelevant that 90 percent of the world's population and almost all states bitterly condemned the
invasion. These nations just don't understand their
own will. Their *actual* will was expressed in Security
Council resolutions with which Iraq didn't fully com-
ply, and so on. Therefore, under the subtle and complex communitarian interpretation, the United States
was using force with the authorization of the Security
Council even though the Security Council denied it.[21]
This is a large part of what the academic profession
does. Academics make up complex, subtle arguments

that are childishly ridiculous but are enveloped in sufficient profundity and footnotes and references to al-
legedly deep thinkers so that you can construct a
framework which has, in some strange universe, a kind
of plausibility.

*The current rhetoric around Iraq is that the country was "lib-
erated."*

If you want to know whether a country was liberated, ask
the population. They should be the ones to decide, not
the intellectuals and politicians of the invading country.
And by about five to one, in Western-run polls, Iraqis say
the country is under occupation. In one of the most remarkable poll results I've seen, Iraqis were asked to name
the foreign head of state they most respected. The leading
answer was Jacques Chirac, the president of France, who
was the symbol of opposition to the invasion of Iraq.
Chirac polled far above Bush. The pathetic Tony Blair
trailed even farther behind. In some of the polls, to my ut-
ter astonishment, a substantial majority of Iraqis say U.S.
forces should leave, which is remarkable given how bad
the security situation is there.[22]

Actually, if you look at the poll results, Iraqis show a
much more sophisticated understanding of the West than

we do of ourselves. It's very common for the victims to understand a system better than the people who are holding the stick. If you want to learn about patriarchal families, you don't ask the father, you ask the mother; then maybe you will learn something. For example, Iraqis were asked in a Western poll, Why do you think the United States entered Iraq? They didn't use the word *invade*. There were some Iraqis who agreed with President Bush and 100 percent of Western commentators. One percent said that the goal of the invasion was to establish democracy. Seventy percent said that the goal was to take over Iraq's resources and to reorganize the Middle East—they agreed with Richard Perle and Paul Wolfowitz. That was the overwhelmingly dominant position. Approximately 50 percent said the United States wants to establish a democracy in Iraq but would not permit the Iraqi government to carry out its own policies without U.S. influence.[23] In other words, they understand that the United States wants democracy if the U.S. can control it. And that's correct. A democracy is a system in which you are free to do whatever you like as long as you do what we tell you. That ought to be taught in elementary schools here. The evidence is so overwhelming that it's boring to repeat it, but American commentators can't understand it. Iraqis, on the other hand, seem to have no trouble understanding it, in part because they know their own history. The British artificially carved out Iraq in 1920, and set the borders so that Britain, not Turkey, would get control of the oil in the north. And they ensured that Iraq would be a dependency by cutting off its access to the sea. That's the point of the British colony of Kuwait. Then the British declared Iraq to be a free, independent country, running its own affairs. If you look at the British Colonial Office records, which were formerly secret but are now public, the British said that Iraq will be a free country but will be governed by an "Arab façade," behind which the British will still rule.[24] Iraqis don't have to read the secret records. They know their own history. They know how free they were.

Furthermore, Iraqis just have to look at what's happening right now. It's kind of striking to see the U.S. media try to get around the fact that while we're so passionately dedicated to democracy, we're also desperately trying to evade Iraqi calls for an election. This is pretty hard to miss. And Iraqis don't have to read the *Washington Post* to discover that the United States is constructing its largest embassy in the world in Baghdad or that Washington is insisting on a status-of-forces agreement in which the sovereign Iraqi government will grant the United States the right to keep as many military troops and bases in Iraq as it wants and for as long as it wants.[25] They don't have to read the business press in the United States to discover that the occupying authorities have imposed an economic regime that no sovereign state would accept for a moment, which completely opens up Iraq to takeover by foreign corporations. They can see that the economic system that is being imposed on them is a Bush administration dream. Iraqi businessmen are screaming about it, because they know they will never be able to compete with other countries under these conditions.[26] The highest tax rate in Iraq is now only 15 percent—so that means no taxes and no constraints on foreign investment. The only sector excluded from complete foreign ownership is oil, because that would have been too blatant. But if you read between the lines, you see Halliburton executives explaining that the work they're doing now, with nice taxpayer subsidies, will put them in a good position to manage and control Iraq's oil resources in the future.[27]

We are now seeing some criticism in the mainstream media of the invasion of Iraq.

The criticism we are seeing, though, does not question the basic assumptions behind the invasion. The criticism

is that the United States is trying to do the right thing but
Bush is doing it badly. Let's go back to Robert McNamara.
When McNamara wrote his book *In Retrospect*, he was
highly praised by humanist doves.[28] They said, we're vin-
dicated: McNamara finally came around and agreed we
were right all along. What did he say? He apologized to
the American people because he didn't tell them soon
enough that the war was going to be costly for Americans, and he's really sorry about this. Did he apologize to
the Vietnamese? There is not one word of apology to the
Vietnamese. We killed a couple million Vietnamese and
destroyed the country. Vietnamese people are still dying
from the chemical warfare that McNamara initiated. But
none of those actions merit an apology. The premises behind the Vietnam War are accepted across the board. We
were trying to defend South Vietnam, but it was costly to
us so we had to stop. Only within that framework can
you have criticism.

The same is true now with the attack on Iraq. The critics of the war point out that Bush didn't tell us the truth
about weapons of mass destruction. Suppose he had told
us the truth. Would it change anything? Or suppose he
had found them. Would that change anything? If you
want to find weapons of destruction, you can find them

all over the place. Take, say, Israel. There is a great concern right now about proliferation of nuclear weapons, as there should be. This morning's *New York Times* has an op-ed by Mohamed El-Baradei, the director-general of the International Atomic Energy Agency (IAEA), which begins by noting that weapons proliferation is increasing, which is an extremely dangerous threat to the world.[29] Yes, it's increasing. Why? There are many reasons, but one of them is that Israel has hundreds of nuclear weapons, as well as chemical and biological weapons, which is not only a threat in itself but encourages others to proliferate in response and in self-defense. Is anybody saying anything about this? Actually, General Lee Butler, the former head of the Strategic Air Command, did acknowledge this problem in a speech a few years ago. He said "it is dangerous in the extreme that in the cauldron of animosities that we call the Middle East, one nation has armed itself, ostensibly, with stockpiles of nuclear weapons, perhaps numbering in the hundreds, and that that inspires other nations to do so."[30] He didn't name the country, but obviously he meant Israel.

Just a few days ago the leading Israeli journal, *Ha'aretz*, in its Hebrew edition—they didn't have it in the English edition—published a very interesting leak from some unidentified military source, which is obscure but would be investigated by anyone concerned with proliferation. The leak said that the United States is providing the Israeli air force with *himush "myuhad"*—"'special' weaponry"—which may very well be a code word for nuclear warheads for the advanced U.S. aircraft that Israel flies.[31] Maybe reporters and commentators here don't want to talk about this subject, but you can bet your life that Iranian intelligence is reading these reports. So how are they going to respond? By proliferation.

If you want to worry about countries with weapons of mass destruction, you don't have to look very far. The United States is itself increasing proliferation by rejecting treaties, by barring any effort to stop militarization of space, by developing what they call "mini nukes," which are actually massively destructive nuclear weapons. In his column, El-Baradei says politely that we should try to implement the treaty to block transmission of materials for developing enriched uranium. He doesn't say, however, that the world has been trying to do this for some time but the Bush administration isn't participating.

Militarization of space alone is an extremely serious problem. UN disarmament commissions have been immobilized for years. This goes back to the Clinton administration's refusal to permit measures that would ban the militarization of space. Right after the announcement, with great fanfare, of the National Security Strategy in September 2002, another announcement was made that received no coverage, even though it may be even more important. The Air Force Space Command, which is in charge of advanced space-age nuclear and other weaponry, released its projection for the next several years, in which it said that the United States is going to move from "control" of space to "ownership" of space.[32] Ownership of space means no potential challenge to U.S. control of space will be tolerated. If anyone challenges us, we'll destroy them.

What does ownership of space mean? It's spelled out in high-level documents, some leaked, some public. It means putting platforms in space for highly destructive weapons, including nuclear and laser weapons, which can be launched instantaneously, without warning, anywhere in the world. It means hypersonic drones that will keep the whole world under photo surveillance, with high-resolution devices that can tell you if a car is driving across the street in Ankara, or whatever you happen to be interested in, meaning the whole world is under surveillance.[33] We probably ultimately won't even need forward bases, because the United States will be able to launch attacks from a command post in the mountains of Colorado or Montana.

How do you think the world will react to this? Russia and China have already reacted with an increase in military spending for offensive military weapons. Russia has shifted its missile system to launch on warning, meaning automated response. Russia's nuclear weapons program was always extremely dangerous, but now with deteriorating command and control systems, it's even more dangerous.[34] Just to give you an indication of how dangerous, in 1995 we came a few minutes from a nuclear war. Russian computerized systems interpreted a scientific rocket launch from Norway as a first strike and went into action. Luckily, Boris Yeltsin called off the attack.[35] Today, Russia's systems are much worse. The Chinese have also reacted. I wouldn't be at all surprised if the Chinese moon shot was a response to U.S. designs on space, intended to convey the message, "We're not going to allow you to own space." And that can have great dangers.

Meanwhile, the United States has assumed a far more aggressive posture. More money is now going into so-called missile defense. Everyone interprets the missile shield as an offensive weapon that is supposed to provide protection against retaliation to a U.S. first strike. And everyone knows how other countries will respond, namely, by increasing their offensive military capacities. The other mode of response is terror. Those are the weapons available to the potential targets of U.S. attack. So we're asking for an increase in terror, an increase in proliferation, an increase in threats to people in the United States. That's the consequence of these programs, and it's not particularly secret. Why do it? For short-term gain. If it leads to long-term disaster, that's somebody else's problem.

The same logic applies in other domains. The concern over global warming has now reached a stage that even the Pentagon is producing studies about the severe threat of global warming within the next twenty or thirty years.[36] One serious prediction is that there could be a fairly sudden shift in the Gulf Stream, which would turn northern Europe into Labrador and Greenland, and might turn large parts of the United States into desert.[37] Rising sea levels could wipe out Bangladesh and kill who knows how many people. The most arable lands in Pakistan may become like the Sahara.[38] The effects of all of this are indescribable. Are we doing anything about it? No. We don't care. Meaning planners don't care. It's not part of their framework. If you're a corporate manager, you don't care about what's going to happen ten years from now. You have to make sure you get your big bonus and stock options next year, not ten years from now. That's somebody else's department. This fanatic ideology is built into the institutional structure. You can't even blame individuals for it, any more than you can blame McNamara for carrying out a cost-benefit analysis that shows how to maximize the number of Japanese civilians you can murder. It's like what Hannah Arendt said about Adolf Eichmann.[39] You do your job. Other considerations aren't part of your domain.

About this short-range vision, these people have children, grandchildren. Aren't they totally dismissing their futures?

Look at our own recent history. Around 1950, the United States had a position of security. There wasn't a threat within shouting distance—except for one *potential* threat: intercontinental ballistic missiles with thermonuclear warheads. They weren't yet available, but they were beginning to be developed. And they would be a threat to the U.S. heartland, could destroy it, in fact. Now, if you care about your children and your grandchildren, wouldn't you do something to prevent that threat from developing? Could anything have been done? Nothing was tried, so we don't know. Surely, at the very least, one could have explored treaties that would have blocked the development of these weapons. In fact, it's not unlikely that the Russians would have agreed to such treaties. They were so far behind technologically, and legitimately frightened and threatened, that they might well have agreed not to develop these weapons. As we know from the newly opened Russian archives, they also understood that the United States was trying to spend them into economic destruction by compelling them to enter an arms race that they couldn't survive—remember, their economy was much smaller than ours. So it's possible, in fact likely, that they would have accepted such a treaty. What's the historical record on this? In the standard magisterial history, McGeorge Bundy, a national security adviser who had access to declassified records, mentions, more or less in passing, that he was unable to find any mention of even the possibility of pursuing this option.[40] It's not that it was suggested and rejected; he says it wasn't mentioned. Did you have to be some kind of a genius to understand in the early 1950s that that was the one potential threat to the United States and that it might destroy your grandchildren? No, you just had to have the intelligence, the knowledge of the world of a normal high school student. These were not stupid people. Dean Acheson, Paul Nitze, George Kennan, and the rest. But it didn't occur to them, because they had higher aims, like maximizing short-term power and privilege.

What do you say to someone reading this interview who says, "These are enormous problems. What can I as an individual do about them?"

There's a lot we can do. We're not going to be thrown into prison and face torture. We're not going to be assassinated. We have enormous privilege and tremendous freedom. That means endless opportunities. After every talk I give in the United States, people come up and say, "I want to change things. What can I do?" I never hear these questions from peasants in southern Colombia, Kurds in southeastern Turkey under miserable repression, or anybody who is suffering. They don't ask what they can do; they tell you what they're doing. Somehow the fact of enormous privilege and freedom carries with it a sense of impotence, which is a strange but striking phenomenon. The fact is, we can do just about anything. There is no difficulty in finding and joining groups that are working hard on issues that concern you. But that's not the answer that people want.

The real question people have, I think, is, "What can I do to bring about an end to these problems that will be quick and easy?" I went to a demonstration, and nothing changed. Fifteen million people marched in the streets on February 15, 2003, and still Bush went to war; it's hopeless. But that's not the way things work. If you want to make changes in the world, you're going to have to be there day after day doing the boring, straightforward work of getting a couple of people interested in an issue, building a slightly bigger organization, carrying out the next move, experiencing frustration, and finally getting somewhere. That's how the world changes. That's how you get rid of slavery, that's how you get women's rights, that's how you get the vote, that's how you get protection for working people. Every gain you can point to came from that kind of effort—not from people going to one demonstration and dropping out when nothing happens or voting once every four years and then going home. It's fine to get a better or maybe less worse candidate in, but that's the beginning, not the end. If you end there, you might as well not vote. Unless you develop an ongoing, living, democratic culture that can compel the candidates, they're not going to do the things you voted for. Pushing a button and then going home is not going to change anything.

HISTORY
AND MEMORY

CAMBRIDGE, MASSACHUSETTS (JUNE 11, 2004)

Tell me about the painting that hangs in your office. It's rather gruesome.

It's a picture of the angel of death standing over the archbishop of El Salvador, Oscar Romero, who was assassinated in 1980.[1] Romero was assassinated only a few days after he had written a letter to President Jimmy Carter pleading with him not to send aid to the military junta in El Salvador, which would be used to crush people struggling for their elementary human rights.[2] The aid was sent, and Romero was assassinated. Then Ronald Reagan took over. The kindest thing you can say about Reagan is that he may not have known what the policies of his administration were, but I'll pretend he did. The Reagan years were a period of devastation and disaster in El Salvador. Maybe seventy thousand people were slaughtered.[3] The decade began with the assassination of the archbishop. It ended, rather symbolically, with the brutal murder of six leading Latin American intellectuals, Jesuit priests, by an elite battalion, trained, armed, and run by the United States, which had a long, bloody trail of murders and massacres behind it.[4] The painting shows the priests, along with their housekeeper and her daughter, who were also murdered. Just about everyone from south of the Rio Grande who comes to visit the office recognizes the image. Almost no one from north of the Rio Grande does.

When enemies commit crimes, they're crimes. In fact, we can exaggerate and lie about them with complete impunity. When we commit crimes, they didn't happen. And you see that very strikingly in the cult of Reagan worship, which was created through a massive propaganda campaign. Reagan's regime was one of murder, brutality, and violence, which devastated a number of countries and probably left two hundred thousand people dead in Latin America, with hundreds of thousands of orphans and widows. But this can't be mentioned here. It didn't happen.

The person responsible for one component of this terror, the Contra war in Nicaragua, was the person known as the "proconsul" of Honduras, John Negroponte. Negroponte was U.S. ambassador to Honduras, which served as the base for the terrorist army attacking Nicaragua. He had two tasks as proconsul. First, to lie to Congress about atrocities carried out by the Honduran security services so that the military aid could continue to flow to Honduras. And second, to supervise the camps in which the mercenary army was being trained, armed, and organized to carry out the atrocities, atrocities for which it was condemned by the World Court. Now Negroponte is the proconsul of Iraq. The *Wall Street Journal*, to its credit, had an article pointing out that Negroponte is going to Iraq as a "modern proconsul" and that he learned his trade in Honduras in the early 1980s.[5] In Honduras, I might add, he was in charge of the biggest CIA station in the world. He's now in charge of the biggest embassy in the world. But all of this didn't happen and it doesn't matter, because we did it. And that's a sufficient reason for effacing it from history.

Today's New York Times *is full of the solemnity and pageantry of a state funeral honoring President Reagan, someone who called the Contras in Nicaragua "the moral equivalent of the founding fathers."[6] In the front-page story "Legacy of Reagan Now Begins the Test of Time," R. W. Apple, Jr., writes about Reagan's "extraordinary political gifts," including "his talents as a communicator, his intuitive understanding of the average American, his unfailing geniality."[7]*

In R. W. Apple's article, which is typical, the entire record of Reaganite atrocities is completely erased. Take Africa, for example. During the Reagan years, the administration had a policy toward South Africa of "constructive engagement." There was strong opposition to apartheid at the time, and Congress had passed legislation banning aid for South Africa. The Reaganites had to find ways to get around congressional legislation in order to in fact increase their trade with South Africa. So they said that South Africa was defending itself against one of the "more notorious terrorist groups" in the world, namely Nelson Mandela's African National Congress.[8] This was a period of massacres, devastation, and destruction, all of which is effaced.

One of the things that happened during Reagan's administration was the invasion of Grenada. You were in Boulder, Colorado, that day, October 25, 1983, and you began your talk by saying, "The latest U.S. intervention as of this morning is Grenada." Reagan said that the building of an airfield in Grenada "can only be seen as Soviet and Cuban power projection into the region."9

Again the kindest thing you can say about Reagan is that he probably didn't know what he was saying. He was handed his notes by speechwriters, including his jokes, incidentally. But, pretending that he knew, the claim was that Grenada was a Soviet-Cuban beachhead because some Cuban contractors, under British planning and authorization, were building an airfield. The Russians, if they could somehow find Grenada on a map, were going to use it as an air base to attack the United States.

Reagan was an incredible coward. Somebody who could believe that an air base in Grenada could be used to attack the United States does not even reach the level of a laughingstock. And the same thing happened with Nicaragua. Reagan declared a national emergency because the government of Nicaragua posed "an unusual and extraordinary threat to the national security and foreign policy of the United States."[10] He then explained that Nicaragua was "a privileged sanctuary for terrorists and subversives just two days' driving time from Harlingen, Texas."[11] Anyone looking at this wouldn't know whether to laugh or cry. In fact, you have to cry, because this was all part of a process of destroying Nicaragua and very seriously harming the United States.

Reagan said he was intervening in Grenada to save the lives of students at St. George's University School of Medicine.

The claim was the United States was protecting American students at the medical school.[12] The fact that Cuba made offers instantly to negotiate the whole issue was suppressed by the media. It was kind of leaked quietly after, when it was too late. And, of course, the real reason for the invasion was not obscure. Just a couple of days before, there had been a bombing in Lebanon in which 240 American marines were killed. And they had to cover this up with a grand gesture defending us from destruction by Grenada. After the invasion, Reagan stood up and said, "Our days of weakness are over. Our military forces are back on their feet and standing tall."[13]

Incidentally, the idea that Reagan struck a chord among the American people is simply not true. He was not a popular president. Even the press sometimes has to concede this now. Take a look at the Gallup polls. Reagan's poll ratings through his years in office were roughly average, below every one of his successors, except for Bush II. By 1992, Reagan had become the most unpopular living former president apart from Richard Nixon.[14] Then came an immense propaganda campaign, which has been going on for about ten years, to turn him into a semi-divinity, which has had some success. If you follow the propaganda campaign and check the polls, you see that the reverence for the imperial leader increased as the propaganda campaign mounted. It's true that people are susceptible to imperial propaganda.

This state funeral today in Washington is intriguing. As the *Times* pointed out, it is following the script of a three-hundred-page funeral plan, which spells out in precise detail what should happen every minute of the imperial ceremony.[15] There has been nothing like it in U.S. history. John F. Kennedy's funeral was totally different; that followed the assassination of a living president. To find anything that compares to this, you would have to go back to the outlandish cult of George Washington that was developed in the early nineteenth century. Washington was turned into the perfect human being, the most amazing creature who ever walked the face of the earth, much like what you might find in North Korea about Kim Il Sung. This was during a period when people were trying to create a unified country out of the separate colonies. Until the Civil War, roughly, the term *United States* was plural, not singular—the *states* that are united. The effort to forge a nation required a major propaganda effort, especially by nineteenth-century standards. But from then until now, there's been nothing comparable to the cult of Reagan.

Your office here in a new building at MIT is opposite another new one that's called the Center for Learning and Memory. One can only speculate as to what goes on there. But I'd like you to talk about memory and knowledge of history as a tool of resistance to propaganda.

It was well understood, long before George Orwell, that memory must be repressed. Not only memory but consciousness of what's happening right in front of you must be repressed, because if the public comes to understand what's being done in its name, it probably won't permit it. That's the main reason for propaganda. Otherwise there is no point in it. Why not just tell the truth? It's easier to tell the truth than to lie. You don't get caught. You don't have to put any effort into it. But power systems never tell the truth, if they can get away with it, because they simply don't trust the public.

On May 27, the *New York Times* ran an article about the interchanges between Henry Kissinger and Richard Nixon that included one of the most incredible sentences I've ever read. Kissinger fought very hard through the courts to try to prevent the transcripts from being released, but the courts permitted it. You read through them, and you find that at one point Nixon informed Kissinger that he wanted to launch a major assault on Cambodia under the pretense of airlifting supplies. He said, "I want them to hit everything." And Kissinger transmitted the order to the Pentagon to carry out a "massive bombing campaign in Cambodia. Anything that flies on anything that moves."[16] That is the most explicit call for what we call genocide when other people do it that I've ever seen in the historical record.

Right at this moment, Slobodan Milošević, the former president of Yugoslavia, is on trial, and prosecutors are somewhat hampered because they can't find direct orders linking Milošević to major atrocities on the ground in Bosnia. Suppose they found a statement from Milošević saying, "Hit everything. Anything that flies on anything that moves." The trial would be over. Milošević would be sent away for multiple life sentences. But they can't find any such document.

Was there any reaction to the Nixon-Kissinger transcripts? Did anybody notice it? Actually, I've brought this comment up in a number of talks, and I've noticed that people don't seem to understand it. They might understand it the minute I say it, but not five minutes later, because it's just too unacceptable. We cannot be people who openly and publicly call for genocide and then carry it out. That can't be. So therefore it didn't happen. And therefore it doesn't even have to be wiped out of history, because it will never enter history.

In your essay "On War Crimes" from At War With Asia, *you cite Bertrand Russell's introduction to the international war crimes tribunal on Vietnam. Russell said, "It is in the nature of imperialism that citizens of the imperial power are always among the last to know—or care—about circumstances in the colonies."*[17]

I disagree with Russell when he says that citizens of the imperial power are the last to care. I think they do care, and I think that's why they're the last to know. They're the last to know because of massive propaganda campaigns that keep them from knowing. Propaganda can be either explicit or silent. When you're silent about your own crimes, that's propaganda, too. And the reason for the propaganda, both kinds, is that people do care, and if they find out what's really happening, they're not going to let it continue. In fact, we actually see that right now. You won't read it in the headlines, but take, say, the recent events in Falluja, Iraq. The marines invaded Falluja, and killed nobody knows how many people, but likely hundreds.[18] We never investigate our own victims, so we don't know the numbers. The United States had to back off and, though no one will say it, effectively conceded defeat. The marines turned the city over to what amounts to the former army of Saddam Hussein. Why did that happen? Suppose there had been an assault like this in the 1960s. It would have been settled very simply with B52s and massive ground operations to wipe the place out. Why didn't the U.S. military do that this time? Because the public won't tolerate it.

In the 1960s, executive power was so extreme that the government could get away with anything. It was just taken for granted that it's our right to massacre and destroy at will. So there was virtually no protest against the Vietnam War for years, and operations like the one in Falluja went on constantly. Not anymore, though. Now the public won't tolerate it. That's one major reason why the United States can no longer carry out the kinds of murderous operations that it was once easily able to carry out.

I spend a lot of time looking at declassified government documents. You take a look at secret documents from the United States or, to the extent that I know about them, other countries. If they are protecting secrets, who are they keeping them from? Mostly the domestic population. A very small proportion of these internal documents have anything to do with security, no matter how broadly you interpret it. They primarily have to do with ensuring that the major enemy—namely, the domestic population—is kept in the dark about the actions of the powerful. And that's because people in power, whether it's business power or government power or doctrinal power, are afraid that people do care, and therefore you have to, as Edward Bernays said, consciously manipulate their attitudes and beliefs.

June 2004 marks the fiftieth anniversary of the U.S. coup overthrowing the democratically elected government of Jacobo Arbenz in Guatemala.[19] Dwight D. Eisenhower, after the coup, said to Allen Dulles and other top officials, "Thanks to all of you. You've averted a Soviet beachhead in our hemisphere."[20] Stephen Schlesinger and Stephen Kinzer wrote a book on the

coup called Bitter Fruit.[21] *Schlesinger in a* Nation *article
called it "one of the blackest episodes in the CIA's history."*[22]
Comment on what happened in Guatemala.

Bitter Fruit is a good book. The coup was not a dark moment in the CIA's history, though. The CIA acted, as it
constantly acts, as an agency of the White House to carry
out actions with what's called "plausible deniability." The
CIA is assigned the responsibility of committing the
crimes and atrocities, and then if anything goes wrong,
you can blame it on "rogue" elements at the agency. But
that's a joke. It's very hard to find a case in which the CIA
acted outside presidential authority. And in the case of
the overthrow of Arbenz, Eisenhower gave the orders. As
to Guatemala being a Soviet beachhead, Eisenhower
knew perfectly well that his administration had been trying very hard to force Guatemala to accept Eastern European arms. Guatemala had a democratic government, to
which the U.S. was strongly opposed. A Guatemalan poet
called this brief interlude "Years of Spring in a country of
eternal tyranny."[23]

After the dictatorship of Jorge Ubico Castañeda was
overthrown in 1944, Guatemala finally had an authentic
democratic government, with enormous popular support

because of its progressive social policies. For the first time, the government mobilized peasants to participate in the political system. A real democracy was developing, which could have influenced other countries in Latin America. The United States considered this an incredible crime. Dulles and Eisenhower, in secret discussions, were profoundly concerned. They worried that Guatemala might be supporting strikes in nearby Honduras or aiding José Figueres, the leading figure of Central American democracy, who was trying to overthrow a dictatorship in Costa Rica. When the United States threatened the country with attack, Guatemala sought military assistance from Europe, which the United States blocked. Finally, Guatemala, trying to defend itself from an attack by the hemispheric superpower, made the tactical mistake of accepting military aid from the only country that would help out, Czechoslovakia. The U.S. government then triumphantly discovered that Czech arms were going to Guatemala, and this fact was trumpeted as a threat to the United States. How can the United States survive if Guatemala has some rifles from Czechoslovakia? This was used as the pretext for the invasion.

Incidentally, although we have an enormous amount of information about Guatemala, what we know is still quite limited. Part of the reason is that the Reaganites, who were not conservatives but extreme statist reactionaries, blocked the regular release of archival records that would have shed more light on this period. Generally, U.S. law requires the State Department to declassify and release records after a thirty-year period. The Reagan administration blocked this because they didn't want the public to know what had happened in Guatemala in 1954 and Iran in 1953.[24] People might learn the truth about what the state was up to, and they wouldn't accept it.

The newspaper of record, the New York Times, *had a role in the 1954 Guatemala coup. The director of the CIA asked the* Times *to keep its correspondent Sydney Gruson away from the story, and the newspaper's publisher, Arthur Hays Sulzberger, complied.*[25]

The *Times* was a cheerleader for the coup in Guatemala and also applauded the coup in Iran in 1953. Thomas McCann, the public relations officer of the United Fruit Company, actually wrote an interesting book about this, *An American Company*, in which he describes the propaganda efforts, led by Edward Bernays, to persuade the public and the press to support the coup. And then he says, "It is difficult to make a convincing case for manipulation of the press when the victims proved so eager for the experience."[26]

The cover of the Pakistani writer and activist Eqbal Ahmad's book Terrorism: Theirs and Ours *has a photograph of Ronald Reagan sitting in the White House with a group of mujahideen from Afghanistan. This is not a photograph that is being widely circulated in any of the major media. The Reagan administration was instrumental in supporting the mujahideen, elements of which later morphed into the Taliban and Al Qaeda.*[27]

They went beyond supporting them. They organized them. They collected radical Islamists from around the world—the most violent, crazed elements they could find—and tried to forge them into a military force in Afghanistan. The mujahideen were armed, trained, and directed by Pakistani intelligence mainly, but under CIA supervision and control, with the support of Britain and other powers. You could argue that this would have been legitimate if it had been for the purpose of defending Afghanistan, but it wasn't. In fact, it probably prolonged

the war in Afghanistan. The Soviet archives suggest
Moscow was ready to pull out of Afghanistan in the early
1980s. But that wasn't the point. The point was not to defend the Afghans but to harm the Russians. The mujahideen carried out terrorist activities right inside Russia.
And these same forces later morphed into what became
Al Qaeda. Incidentally, those terrorist activities stopped
after the Russians pulled out of Afghanistan, because
what the mujahideen were trying to do is just what they
said: to protect Muslim lands from "the infidels."

Actually, Al Qaeda, if you look back, was barely
mentioned in U.S. intelligence reports until 1998. Clinton's bombing of Sudan and Afghanistan in 1998 effectively created Al Qaeda, both as a known entity in the
intelligence world and also in the Muslim world. In fact,
the bombings created Osama bin Laden as a major symbol, led to a very sharp increase in recruitment and financing for Al Qaeda–style networks, and tightened
relations between bin Laden and the Taliban, which previously had been quite hostile to him. The bombing of
Sudan, in particular, infuriated people throughout the
Arab world. It's another moment in history that didn't
happen because we did it. The United States knew perfectly well that it was targeting a major producer of

pharmaceutical and veterinary supplies for a poor African country. Of course, that's going to have devastating effects. Just how devastating we don't know because, again, we don't investigate or care about the results of our crimes. But the few credible estimates that are available, one from the German ambassador published in the ultraleft *Harvard International Review* and another in the *Boston Globe*, plausibly estimate several tens of thousands of deaths as a consequence of the bombing—maybe more, maybe less.[28] Here, that's not an issue. If Al Qaeda blew up half the pharmaceutical supplies in some country that mattered—the United States or England or Israel—we wouldn't say, "Oh, well, it's no big deal." But when we did it, it didn't happen, and the consequences didn't occur. And if anybody even dares to mention this, it just leads to hopeless tantrums, because you're not allowed even to mention the fact that the United States can just thoughtlessly carry out major crimes.

Osama bin Laden himself only became anti-American around 1991, for several reasons. The United States and Saudi Arabia refused to allow him to carry out a jihad against Saddam Hussein during the first Gulf War. But the main reason was that the United

States had bases in Saudi Arabia, near two of the holiest cities in Islam.

I interviewed Eqbal Ahmad in August 1998, a couple of weeks after Clinton launched cruise missile attacks on Afghanistan and Sudan, and he said, "Osama bin Laden is a sign of things to come. . . . The United States has sowed in the Middle East and in South Asia very poisonous seeds. These seeds are growing now. Some have ripened, and others are ripening. An examination of why they were sown, what has grown, and how they should be reaped is needed. Missiles won't solve the problem."[29]

That's a very perceptive statement. And, in fact, by now there is quite good analytic literature on how these seeds have developed. The best book on this topic is *Al-Qaeda*, by British investigator Jason Burke, which confirms what Eqbal Ahmad predicted.[30] Burke argues that Al Qaeda is not an organization but a network of very loosely affiliated and mostly independent organizations with a similar ideology, a "network of networks." According to Richard Clarke's book *Against All Enemies*, U.S. intelligence paid no special attention to Al Qaeda or Osama bin Laden until 1998. In fact, they didn't even use the term *Al Qaeda*.[31] But, as Eqbal predicted, the bombings in Sudan and Afghanistan led to Al Qaeda and bin Laden becoming major symbols. These attacks, along with the invasion of Afghanistan, led to a big increase in recruitment and financing for Al Qaeda–type groups. Burke rightly says, "Every use of force is another small victory for bin Laden," helping him mobilize the constituency he hopes will see the West as crusaders trying to destroy the Muslim world.[32] The war in Iraq had exactly the same effect.

Just this morning the State Department admitted that, as they politely put it, they were "wrong"—in other words, lying outright—when they claimed in their report on "Patterns in Global Terrorism" that terror had been reduced thanks to Bush.[33] In fact, it had increased, they now concede quietly, though that's been known for a while.[34] Part of the increase is due to the war in Iraq, which was totally predictable. In fact, intelligence agencies and analysts predicted that if the U.S. invaded Iraq it would increase terrorism, for pretty obvious reasons.[35]

There is an odd charade going on now in the intellectual world and in Washington based on the so-called revelations of Richard Clarke, Paul O'Neill, the former treasury secretary, and others that the neoconservatives in the Bush II administration ranked invading Iraq higher than the war on terror. The only thing surprising about these revelations is that anybody is surprised. How can you be surprised? They invaded Iraq, after all, knowing that it was very likely to increase the threat of terror. That demonstrates what their priorities are. End of story. Furthermore, they're perfectly reasonable priorities from their point of view. They don't care that much about terror. What they care about, as Chalmers Johnson rightly points out, is having military bases in a dependent client state right at the heart of the world's largest oil-producing region. That's important. Not because the United States wants the actual oil—it's going to get oil one way or another on the market—but because it wants to *control* the oil, which is a totally different matter. It has been understood since the 1940s that control of the oil is a major lever against your enemies. And the U.S. enemies are Europe and Asia. Those are the regions of the world that could move toward independence. One of the ways to prevent that is to keep your hand on the spigot.

Every four years, U.S. voters are faced with a choice between what has been called the "lesser of two evils." You've said that there is "a fraction" of difference in this upcoming election between George Bush and John Kerry, which has raised some eyebrows. Could you expand on your position?

There are differences. Kerry and Bush have different constituencies, and have different groups of people around them. On international affairs, I wouldn't expect any major policy changes if Kerry were elected. It would probably be more like the Clinton years, when you had more or less the same policies but more modulated, not so brazen and aggressive, less violent. But on domestic issues there could be some fairly significant differences in outcomes. The people around Bush are real fanatics. They're quite open. They're not hiding it; you can't accuse them of that. They want to destroy the whole array of progressive achievements of the past century. They've already more or less gotten rid of the progressive income tax. They're trying to destroy the limited medical care system. They're going after Social Security. They'll probably go after schools. They don't want a small government any more than Reagan did. They want a huge, massively intrusive government, but one that works for them. They hate free markets. The Kerry people will do something not fantastically different, but they have a different constituency to appeal to, and are much more likely to protect some limited form of benefits for the general population.

There are other differences. A large part of the popular constituency of the Bush people is the extremist fundamentalist religious sector in the country, which is huge. There is nothing like it in any other industrial country.

And Bush has to keep throwing these people red meat to keep them in line. While they're getting shafted by Bush's economic and social policies, he's got to make them think he's doing something for them. But throwing red meat to that constituency is very dangerous for the world, because it means violence and aggression, but also for the country, because it means seriously harming civil liberties. Of course, the Kerry people don't really have that constituency. They would like to have it, but they're never going to appeal to it much. They have to appeal somehow to working people, women, minorities, and others.

These may not look like huge differences, but they translate into quite big effects for the lives of people. Anyone who says, "I don't care if Bush gets elected" is basically telling poor and working people in the country, "I don't care if your lives are destroyed. I don't care whether you are going to have a little money to help your disabled mother. I just don't care, because from my elevated point of view I don't see much difference between the two candidates." That's a way of saying, "Pay no attention to me, because I don't care about you." Apart from its being wrong, it's a recipe for disaster if you're hoping ever to develop a popular movement and a political alternative.

THE DOCTRINE OF GOOD INTENTIONS

CAMBRIDGE, MASSACHUSETTS (NOVEMBER 30, 2004)

You've written about the "doctrine of good intentions." Occasionally U.S. policy is marred by the proverbial "bad apples" and "tragic mistakes," but basically the record of our goodness continues unimpeded.

The standard story in scholarship and in the media is that there are two conflicting tendencies in U.S. foreign policy. One is what's called Wilsonian idealism, which is based on noble intentions. The other is sober realism, which says that we have to realize the limitations of our good intentions. Sometimes our noble intentions can't be properly fulfilled in the real world. Those are the only two options.

You see this not just in the United States. Take England. Probably the best newspaper in the world is the *Financial Times* in London. The *Financial Times* printed a column a few days ago by one of their leading columnists, Philip Stephens, that was quite critical of U.S. policy. The problem, he says, is that U.S. strategy is overly dominated by Wilsonian idealism. You need a few drops of sober, "hardheaded realism" to temper this passionate dedication to democracy and freedom.[1]

And Stephens goes on to say that there can no longer be any doubt that George Bush and Tony Blair are motivated by their vision and faith in democracy and rights. We know this because they've said so, and that proves it. But we have to be more realistic and acknowledge that, although Bush and Blair are dedicated to what the press calls "the Bush messianic mission to graft democracy onto the rest of the world." We must understand that Iraqis and others in the Middle East may not be able to rise to the heights that we have planned for them.[2]

As the pretexts for the invasion of Iraq have collapsed—no weapons of mass destruction, no Al Qaeda tie to Iraq, no connection between Iraq and 9/11—Bush's speechwriters had to conjure up something new. So they conjured up his messianic vision to bring democracy to the Middle East. When Bush gave his speech announcing his new vision, the leading commentator at the *Washington Post*, David Ignatius, a respected editor and correspondent, just fell over in awe. He described the Iraq war as perhaps "the most idealistic war fought in modern times—a war whose only coherent rationale, for all the misleading hype about weapons of mass destruction and al Qaeda terrorists, is that it toppled a tyrant and created the possibility of a democratic future." This vision of a "democratic future" is led, according to Ignatius, by the "idealist in chief," Paul Wolfowitz, who has probably the most extreme record of passionate hatred of democracy of anybody in the administration. But it doesn't matter. The proof is that Ignatius was with Wolfowitz when he went to the town of Hilla and spoke to Iraqis about Alexis de Tocqueville.[3] Hilla also happens to be the town where the first major U.S. massacre of Iraqis during the invasion took place, but put that aside, as well.[4]

Ignatius represents one side of the spectrum. Then you go to the other side of the spectrum, the critics who say that the vision is noble, inspiring, but we have to be more realistic, face the fact that it's beyond our reach, that Iraqi culture is deficient, and so on. Is there anything new about this debate? Nothing at all. In fact, you would have to work hard to find a counterfactual historical example. The French were carrying out a "civilizing mission," Mussolini was nobly uplifting the Ethiopians. If we had records from Genghis Khan when he was massacring tens of millions of people, he probably also had a "noble vision." See if you can find an exception.

In Deterring Democracy, *you quote Winston Churchill, speaking to Joseph Stalin in Tehran in 1943. Churchill said that "the government of the world must be entrusted to satisfied nations, who wished nothing more for themselves than what they had. If the world-government were in the hands of hungry nations, there would always be danger. But none of us had any reason to seek for anything more. The peace would be kept by peoples who lived in their own way and were not ambitious. Our power placed us above the rest. We were like rich men dwelling at peace within their habitations."[5]*

Churchill is one of the rare exceptions who doesn't only gush about his noble vision, but occasionally tells the truth. Right before the First World War, Churchill argued that Britain must greatly expand its military expenditures to maintain its empire. With his typical eloquence, he said, "We are not a young people with innocent record and a scanty inheritance. We have engrossed to ourselves

[an] altogether disproportionate share of wealth and traffic of the world. We have got all we want in territory, and our claim to be left in the unmolested enjoyment of vast and splendid possessions, mainly acquired by violence, largely maintained by force, often seem less reasonable to others than to us."[6] Those were Churchill's words in a speech to parliament in 1914, later discovered by one of his biographers, Clive Ponting. Churchill published the speech, about twenty years later, but he cut out all the offending statements.

The original cover of your book At War With Asia, *first published by Pantheon and recently reissued by AK Press, has a remarkable black-and-white photograph of a U.S. soldier.*[7]

A soldier with a rope pulling a skinny, half-naked Vietnamese captive behind him.

Fast-forward to Lynndie England in Iraq.

The only difference is that Lynndie England is not a big, beefy soldier, but otherwise it's the same. In fact, you go back to paintings of the conquest of Massachusetts, where we are sitting now, and it's also the same. Go back to the ugliest periods of history, and it's the same. That's a universal image of unconstrained power being exercised over helpless victims. Nobody anywhere near the mainstream could be more critical than John King Fairbank, an opponent of the war and the dean of Asian scholarship. He said that the United States entered Vietnam "through an excess of righteousness and disinterested benevolence."[8] If we had only had more people who had studied Chinese at Harvard, they would have told us that our flood of magnanimous benevolence wouldn't succeed in Vietnam. That's from the extreme left. Or take Anthony Lewis of the *New York Times*, who called the war in Vietnam "a dangerous mistake," which marred our "blundering efforts to do good."[9] That phrase comes out like boilerplate.

In a front-page story in the New York Times, *"Shadow of Vietnam Falls Over Iraq River Raids," John F. Burns wrote that Vietnam "is rarely mentioned among the American troops in Iraq, many of whom were not yet born when the last American combat units withdrew from Vietnam more than 30 years ago. A war that America did not win is considered a bad talisman among those men and women, who privately admit to fears that this war could be lost."[10]*

First of all, I'm one of the few people who don't agree that the United States lost the war in Vietnam. The

United States didn't win its maximal objectives, but it did achieve its major objectives—a substantial victory. There is no way for a huge, powerful state to lose a war against a defenseless enemy. It just can't happen.

A major concern in the late 1940s right through to when Kennedy launched the full-scale war was that an independent Vietnam could be a successful example to its neighbors, such as Thailand and Indonesia, which had major resources, unlike Vietnam. By the mid-1960s, though, South Vietnam, which was the main target of U.S. intervention, had been virtually destroyed, and the chances that Vietnam would ever be a model for anything had essentially disappeared. As Bernard Fall, the respected military historian and Vietnam specialist, put it in 1967, there was every possibility that Vietnam would become "extinct" as a cultural and historical entity.[11]

I don't usually watch television, but I was in a hotel a few months ago and I watched something on CNN about our "Vietnam obsession."[12] The deep thinkers on the show were talking about how the whole presidential campaign was overwhelmed by discussion of Vietnam. The fact is, Vietnam never even entered the campaign. Did anybody ever refer to what had actually happened there? Did anybody ever ask what John Kerry was doing in Vietnam seven years after Kennedy started bombing it, using chemical warfare, and driving the population out, two years after Bernard Fall thought Vietnam might become extinct? Nobody discussed that because the story has to be that we are benevolent, we made a mistake, and we lost because we didn't achieve our maximal goals. Anything outside of that is just unintelligible to an educated person. So Vietnam is an obsession, but only if we totally ignore the Vietnam War.

In fact, by now it's gotten to the point that the *New York Times* is publishing on the front page photographs and accounts of major U.S. war crimes.

Are you referring to the November 8, 2004, issue of the New York Times, *which showed U.S. troops occupying a hospital in Falluja?*[13]

Yes. One of the first acts in the conquest of Falluja was to take over the general hospital, which was a major war crime. And they gave a reason. The reason is the hospital was a "center of propaganda against allied forces" because it was producing "inflated civilian casualty figures."[14] First of all, how do we know they were inflated? Because our dear leader said so. Secondly, the idea that you take over a hospital because it's publishing casualty figures is obscene. The Geneva Conventions could not be more clear. The wording says explicitly and clearly that "Medical and religious personnel shall be respected and protected and shall be granted all available help for the performance of their duties. . . . Medical units and transports shall be respected and protected at all times and shall not be the object of attack."[15] In the attack on Falluja General Hospital, patients were kicked out of their beds and doctors and patients were forced to lie on the floor, handcuffed. This is a grave breach of the Geneva Conventions. In fact the entire political leadership should face the death penalty under U.S. law for these actions. They're all eligible for the death penalty, according to the War Crimes Act passed by the 1996 Republican Congress.[16]

Remember the Russian assaults on Grozny in Chechnya in 1999? Grozny is a city of about the same size as Falluja, three hundred thousand to four hundred thousand people. They bombed it into dust and destroyed it. The Russian assault on Grozny was considered a major war crime, rightly. But when we do the same thing to Falluja, it's liberation. The embedded journalists are talking about the suffering of the marines, who are so hot and are being fired on all the time. I can't imagine that the Russian press or, for that matter, the Nazi press was any worse.

*The Lancet, a respected British medical journal, did some re-
search on deaths in Iraq since the U.S. invasion and came up
with some rather startling numbers that didn't seem to catch
the attention of the mainstream press.*

The Lancet did a careful study, which estimated conserva-
tively that the most probable number of "excess deaths"
due to the war is about one hundred thousand.[17] Their
cluster sample excluded Falluja, where the number of violent deaths was much higher and would have greatly inflated the total; and it included the Kurdish regions,
where there was almost no fighting and which therefore
lowered the national average. So their estimate is probably on the low side. The report was mentioned in the
U.S. media but mostly dismissed, even though it followed
standard techniques of epidemiological studies. In
Britain, the report caused a little more protest, and the
government was forced to produce some utterly idiotic
comments. Tony Blair's spokesperson said that the
study isn't worth anything because "the findings were
based on extrapolation," like every other epidemiological study.[18] And besides, the Iraqi ministry of health—
that is, the ministry of the U.S.-British–imposed client
government—gives a much lower figure.[19] In England at

least they had to discuss it. In the United States, it didn't matter.

Is this new? In the case of Vietnam, we literally do not know within *millions* the real number of civilian casualties. The official estimates are around two million, but the real number is probably around four million. As far as I know, there's been only one public-opinion study in the United States that asked people to estimate the number of Vietnamese casualties from the war. The mean answer was a hundred thousand, about 5 percent of the official figure.[20] It's as if in Germany you asked people how many Jews were killed in the Second World War and they said three hundred thousand. We would think there was a big problem in Germany if that's what Germans were thinking.

How many victims of chemical warfare were there after 1962, when Kennedy started to destroy food crops and ground cover so that there wouldn't be any indigenous support for guerrillas, using dioxin, one of the most carcinogenic elements on earth? There has been an intensive study of the effect of Agent Orange on American troops. At first the Pentagon denied there was any harmful impact from Agent Orange on U.S. troops, but now they accept the findings. But what about the Vietnamese

NOAM CHOMSKY

people, who were being dosed with it? There was a major study in Canada by Hatfield Consultants, and, in fact, some leading U.S. public health figures at various universities have investigated the topic.[21] Exposure to dioxin is correlated closely with cancers and with other horrors, including children being born without arms and brains. Nobody really knows the numbers, but the rough estimates are at maybe half a million or a million Vietnamese died just from chemical warfare.

In Vietnam, you have a striking test of the effects of dioxin, because Agent Orange was used only in the south. The people have the same genes in the north. Hanoi's hospitals are not full of jars with deformed fetuses; Saigon's hospitals are. Actually, Barbara Crossette wrote an article about a decade ago in the *New York Times* noting that "Vietnam is a good place to study. . . . It furnishes an extensive control group," people in the north who weren't sprayed with dioxins.[22] We could learn a lot that would be useful for ourselves if we did a serious study of the difference between the birth deformities and cancer rates in South and North Vietnam. That's the only question that comes up: Can we learn something about our crimes that would be useful to us? Nothing else.

If you take a look at Japanese literature today, a number of new books have come out, detailed scholarly books, with tons of footnotes, that deny there was a massacre in Nanking.[23] Only a couple hundred thousand people were slaughtered. But the Japanese were defeated, so this interpretation is not the standard line. It's kind of a marginal interpretation, which many people reject. And the Japanese are condemned for it.

There are reports that civilians trying to flee Falluja were turned back by U.S. forces and that Iraqi Red Crescent vehicles attempting to deliver medical supplies to besieged and wounded Iraqis were also turned back.[24]

If civilians managed to flee Falluja, they were allowed out—except for men. Men of roughly military age were turned back. That's what happened in Srebrenica in 1995. The only difference is the United States bombed the Iraqis out of the city, they didn't truck them out. Women and children were allowed to leave; men were stopped, if they were found, and sent back. They were supposed to be killed. That's universally called genocide, when the Serbs do it. When we do it, it's liberation.

The New York Times *ran a small article recently by Michael Janofsky titled "Rights Experts See Possibility of a War Crime."*

It says, "Human rights experts said Friday that American soldiers might have committed a war crime on Thursday when they sent fleeing Iraqi civilians back into Falluja. Citing several articles of the Geneva Conventions, the experts said recognized laws of war require military forces to protect civilians as refugees and forbid returning them to a combat zone." And Janofsky quotes a defense department spokesperson who says, "Our forces over there are not haphazardly operating indiscriminately, targeting individuals or civilians. The rules of engagement are researched and vetted, and our forces closely follow them."[25]

It is interesting that one of the only war crimes that the media are talking about is the case of the marine who kind of lost it in the middle of combat and killed a wounded Iraqi.[26] How could Americans sink to such depths? Yes, what he did is a crime, absolutely, but it's a minuscule footnote. In the history of the Second World War, it wouldn't even be mentioned, it's so minor. But here we blow it up as a way of suppressing the real crimes, just as people did with My Lai. My Lai was a minor footnote to the war in Vietnam. It was part of a major military operation, Operation Wheeler—which was directed by guys just like us, in ties and jackets, sitting in air-conditioned offices and targeting B-52 raids on villages. This was one of many operations that killed who knows how many people. But in one particular spot, some uneducated poor GIs in the field, who were scared out of their wits, lost it and killed a couple hundred people. That's the crime. And the criterion is that they're not like us. You get poor, uneducated people who are in the midst of conflict and have every reason to be scared. If they commit a crime, that's horrible. If nice, educated folk like us, sitting in comfort and protection, commit massive crimes—in particular, ordering these crimes—that doesn't matter. By contrast, Nuremberg worked the opposite way. The prosecution didn't go after the soldiers in the field; it went after the civilian commanders.

The Toledo Blade *produced a remarkable, Pulitzer Prize–winning study of Tiger Force, a platoon created as part of the 101st Air-borne, which in 1967 was sent to the central highlands and committed one atrocity after another. It makes for chilling reading.*[27]

It's missing the point, however. Yes, these soldiers committed atrocities. But 1967 was the year when Bernard Fall published his conclusions that "Vietnam as a cultural and historic entity ... is threatened with extinction ... [as] the countryside literally dies under the blows of the largest military machine ever unleashed on an area of this size."[28] Compare the crimes. Yes, what Tiger Force did is atrocious. But what about the guys from Harvard and MIT who planned these attacks and other actions that threatened the extinction of the country? There's no comparison.

Actually, I wrote a chapter in *At War With Asia* on this topic, "After Pinkville," which is the name people first used for My Lai.[29] I originally was asked to write the essay for *The New York Review of Books*—I was still writing for them then—and I agreed only on condition that I would barely mention My Lai.[30] The essay is about the other, much worse, crimes taking place in Vietnam, directed right from Washington. The planners in Washington are the real war criminals, not the soldiers in the field. The chain of command starts with the civilians sitting in Washington. Those were the people who were charged at Nuremberg and at Tokyo. And if we were willing to be even minimally honest, that's who would be charged here, along with everybody who writes about our benevolence and benign intentions, trying to cover up these crimes.

I was recently listening to a recording of your appearance on Firing Line *with William F. Buckley in April 1969. Talking about Vietnam, you said, "A terrifying aspect of our society and other societies is the equanimity and the detachment with which sane, reasonable, sensible people can observe such events, as in Vietnam. I think that's more terrifying than the occasional Hitler, LeMay, or other that crops up. These people would not be able to operate were it not for this apathy and equanimity."*[31]

Which you find mostly among educated people. The general population tends to be quite different.

Why are you putting so much of the onus on the educated class?

Because responsibility correlates with privilege. If you're more privileged, you're more responsible. Take Germany again, the Nazis, maybe the worst period in history. Some poor guy who was sent out to the eastern front and carried out atrocities—he didn't have any choices. If he objected, he would be slaughtered by the command. But Martin Heidegger had choices. He didn't have to write books and articles giving complicated, elaborate supports for the Nazis.

The people who are sitting in places like MIT have choices. They have privilege, they have education, they have training. That carries responsibility. Somebody who is working fifty hours a week to put food on the table and comes back exhausted at night and turns on the tube has many fewer choices. Technically, this person has choices, but they're much harder to exercise, and therefore he has less responsibility. That's just elementary. The people with the privilege and the education and the training are also the decision makers, either in the government or in business or the doctrinal institutions. So, yes, they're the ones responsible, far more than those who don't have any choices.

You are not in favor of an all-volunteer army. Why not?

I was very active in organizing resistance to the Vietnam War in the 1960s. The only reason I escaped a long jail sentence was because the government called off the trials I was caught up in when the Tet offensive took place. But I was never against the draft, and I'm not against it now. If there is going to be an army, I think it should be a citizens' army, not a mercenary army. There are several kinds of mercenary armies. One model is the French Foreign Legion or the Gurkhas, where the imperial power just organizes a mercenary army. Another model is a volunteer army, which is in effect a mercenary army of the disadvantaged. People like us, except for the occasional maniacs, don't volunteer for it. But people like Lynndie England do volunteer, because they come from a background where that's their only opportunity. So you end up getting a mercenary army of the disadvantaged, and that's much more dangerous than a citizens' army.

But it was a citizens' army in Vietnam.

Take a look at the history of Vietnam. The U.S. command committed a major error. It used a citizens' army to fight a vicious and brutal colonial war. And that can work for a while, but not for very long. Right around that time, soldiers started disobeying orders, fragging their officers, getting doped up. The army was falling apart. That's part of the reason why the top brass wanted them out. Top military analysts at the time from inside the Pentagon were saying that we've got to get that army out of there or else we're not going to have an army. It's just collapsing from within.[32]

A citizens' army has ties to the citizen culture. In the late 1960s, for example, during the Vietnam War, a kind of rebellious culture in many respects and civilizing culture in many respects spilled over into the military, and it helped undermine the military, which is a very good thing. That's why no imperial power had ever used the citizens' army to fight an imperial war. If you take a look at the British in India, the French in West Africa, or South Africans in Angola, they essentially relied on mercenaries, which makes sense. Mercenaries are trained killers, but people who are too close to the civilian society are not really going to be good at killing people.

Getting back to the educated class, how did their opinion of the war differ from that of the general population?

By about 1969, around 70 percent of the population in the United States described the war as "fundamentally wrong and immoral," not as a "mistake."[33] And that's about the time when, at the extreme, critical end, people like Anthony Lewis were beginning to whisper timidly that the war was a mistake.

This gap in public and elite attitudes continues through to the most recent polls on a range of issues. In fact, major polls came out from the most prestigious polling organizations in the country, the Chicago Council on Foreign Relations and the Program on International Policy Attitudes at the University of Maryland, right before the November 2004 election, and the results were so astonishing the press couldn't even report them. The polls showed that a large majority of the population is in favor of signing the Kyoto protocol, accepting the International Criminal Court, and relying on the UN to take the lead in international crises. A majority is even in favor of forgoing the Security Council veto when it comes to what's called preemptive war, which is now interpreted as the right of aggression.[34] In other words, the population is very strongly opposed to the bipartisan consensus on preemptive war. Both parties are in favor of it. Articulate opinion is almost entirely in favor of it, with various qualifications: you've got to make sure it doesn't cost too much, and so on. But a large majority of the population is against it, and takes the position that you're only allowed to use force under the terms of the UN Charter. Most people have probably never heard of the UN Charter, but their answers to polling questions reflect pretty much the standard, narrow interpretation of the charter, which says that you can only use force if you're attacked, or if there is an imminent threat of attack, like planes flying across the Atlantic to bomb the United States.

When you get to the Iraq war, the poll results are quite interesting. About 75 percent say the United States should not have attacked Iraq if it did not have weapons of mass destruction or ties to Al Qaeda. Yet roughly 50 percent say that we should have attacked Iraq. And that's after the Iraq Survey Group report showed that there were no weapons of mass destruction or programs and that there were no ties to Al Qaeda.[35] How do you account for this contradiction? Essentially, people believe the propaganda, even after it was disproved. There has been enough of a barrage of government-media propaganda that about half the population still believed that Iraq had weapons of mass destruction or were developing them. A high percentage still think Iraq was tied to Al Qaeda and 9/11.[36] So, yes, they support the war, even though they're generally opposed to war unless we're under imminent threat of attack.

In fact, if you look at interviews with people like Lynndie England, the torturers of Abu Ghraib, and so on, most of them say that they were taking revenge for 9/11. They did it to us. Why shouldn't we do it to them? If you have any degree of privilege and education, you understand that this makes no sense at all. But people who are being driven into the mercenary army by social and economic conditions don't know that. For them one raghead is the same as another raghead. We can talk about how terrible they are—"Look at these uneducated slobs"—but we have no right to do that. We should be talking about ourselves. We are the ones who are inducing people to have those beliefs, either by our silence or by our apathy or our evasions, or often by direct instruction.

Incidentally, on the domestic front, an overwhelming majority of the population, around 80 percent, are in favor of increased health care; around 70 percent want increased aid to education and Social Security.[37] Both parties are opposed. The health care figures are particularly interesting. Pollsters rarely ask people what kind of health care they want, but when they do ask, it usually turns out that either a plurality or a very large majority is in favor of some kind of universal health care. On October 31, a couple of days before the election, the *New York Times* had an article about health care. It said that Kerry was unable to mention any government program that might improve health care because there is so little political support for it.[38] Only maybe three-fourths of the population. But that's the standard reaction. If national health care is ever mentioned, it's called "politically impossible." It has no political support, only the support of most of the population. That tells you what's going on. "Political support" means support of the insurance industry, Wall Street, HMOs, the pharmaceutical industry. That's political support. In fact, if 98 percent of the population wanted universal health care, that would still not be political support.

What all these polls basically show is that the whole population is so far to the left of both parties that you can understand why the polls aren't published. In fact, one of the only honest reports I saw on the Chicago Council on Foreign Relations poll was in *Newsweek*.[39] If you went on and asked people questions, like "What do you think the general mood of the country is?"—I'm sure most people would say, "I'm the only person who believes this. I'm crazy." They never hear any reinforcement for their views in popular discussion; or in either party platform, or in the media.

SEVEN

INTELLECTUAL SELF-DEFENSE

CAMBRIDGE, MASSACHUSETTS (DECEMBER 3, 2004)

You've said that much of the media analysis you do is simply clerical work.

The hidden truth is that a large amount of scholarship is clerical work. In fact, a good deal of science is detailed, routine work. I'm not saying it's easy—you have to know what you're looking for and so on—but it's not an enormous intellectual challenge. There are aspects of inquiry that are serious intellectual challenges, but usually not those concerned with human affairs. There you have to be sensible and self-critical, but anybody can do this work if they want to do it.

For example, driving in this morning, I was listening to BBC, which is about the only program I can tolerate on the radio, and the news reporter mentioned the bombing of a police station in Iraq. She started her report by saying that the problem in Iraq is that the occupation cannot end unless the Iraqi police are capable of providing security there. Just think about that sentence.[1] Suppose the Nazis in France had said, "The occupation can't end unless the Vichy forces are capable of controlling the country." Wouldn't we think there was something very odd about that? The occupation can end this instant. It's a question of what the Iraqi people want. It should have nothing to do with what Britain and the United States want, any more than the occupation of France should have had anything to do with what the Germans wanted. If the police that were being trained by the Germans to run France under their supervision couldn't control the partisans, does that mean the German army can't leave? There's another way of looking at it, which I think happens to be legitimate. But quite apart from whether it's legitimate or not, it's a point of view that cannot even be considered. We must take the standpoint of the occupying armies, whose governments we speak for unquestioningly. There aren't many polls in Iraq, but the few polls there are indicate that a majority of Iraqis want the occupying troops to leave.[2] Suppose that's true. Do we still believe that the occupation can't end until the Iraqi police can control the country, as the BBC simply presupposes without question? Only if you have so deeply absorbed the doctrines of the people with the whip in their hands is this an assumption that is so obvious you can't even question it. Those are the kinds of issues that interest me personally.

You mean finding and decoding those internalized assumptions, like the idea that the United States has the right to invade and conquer any country and to institute an economic system and a government of its choice?

Yes. That's just taken for granted among the educated population. If we can believe the careful and reputable opinion studies that are carried out in the United States, this isn't true, incidentally, for the general U.S. population. Their view, by a substantial majority, is that the United States should leave Iraq if Iraqis want them to leave. A large majority of the population thinks the United Nations, not the United States, should be taking the lead in international crises in general and should be leading reconstruction in Iraq.[3]

My own personal interest, incidentally, is not the media per se but the intellectual culture. The media happen to be the easiest part of the intellectual culture to study. The elite media—the BBC, the *New York Times*, the *Washington Post*, and so on—are the day-to-day expression of the elite intellectual culture and therefore are much easier to study than intellectual scholarship. You can do that, too, but it requires more complex research. In the media you can fairly easily find systematic biases about what's permitted, what's not permitted, what's stressed, what isn't stressed.

Take this morning's *New York Times*, which has an article reporting the views of Gregory Mankiw, the chair of the President's Council of Economic Advisers. Mankiw's a very distinguished and competent technical economist, a highly regarded professor at Harvard in the economics department and the author of one of the main textbooks in the field. So he's speaking from the peak of the economics profession and he's warning, in proper academic tones, that Social Security benefits will have to be reduced because the U.S. government won't have the money to pay for them. This is reported religiously, with the state-

ment that the Social Security system is headed toward fiscal collapse by 2042 "if no changes are made to the current law."[4] We have to make radical changes, preferably
privatize it.

But there is another way of describing the situation:
the Social Security system is not in crisis and will function as it's now set up for at least thirty years, and by
other government estimates about twenty years beyond
that. Social Security is facing a long-term technical problem that can be easily overcome.

Let's assume that there will be a fiscal problem with
Social Security in forty or fifty years. What can we do
about it? There are some easy solutions that are rarely discussed. For example, the Social Security payroll tax is
highly regressive. Any income you make above roughly
$90,000 is not taxed, which means rich and privileged
people are getting a free ride. Is that a law of nature, that
a small percentage of rich people should get a free ride? If
you simply eliminated the cap, there wouldn't be a Social
Security financing problem for years to come.

The people screaming about the Social Security "crisis" also point out that the proportion of working people
to retired people is declining, which means today's working people are going to have to support a growing number

of retired people. That happens to be true, but it's irrelevant. The real number we need to look at is what's called the total dependency ratio, the proportion of working people to the total number of people, not just retirees.

So take, say, the famous baby boomers. How are we going to pay for their retirement? Who paid for them when they were newborns until they were twenty? You had to care for them just as much as you have to care for your aged mother. If you look back at the 1960s, when this generation was coming of age, in fact, there was a huge increase in funding for schools and other programs for children, at a time when the government had less income than it has today. If you could take care of the baby boomers when they were children, why can't you take care of them when they are over sixty? It's not a bigger problem. The problem is manufactured. It's just a question of financial priorities. In fact, because the United States is now a much richer country than it was in the 1960s, it should be easier to take care of these people.

So the proper reporting of this article should be that a distinguished Harvard economist is giving a radically ideological interpretation that may express his personal biases or some other pressures, but doesn't have much to do with the issue. The system is not heading toward disaster. And to the extent that there is a problem with Social Security, there are a variety of ways of dealing with it. A serious journalist would go on to ask, "What's behind the drive to destroy Social Security?" It's quite transparent. The leading "solution" to the Social Security "crisis" is private investment accounts. Instead of a highly efficient government system, with very low administrative costs, we're moving toward a system with very substantial administrative costs, but costs that will be transferred to the right pockets, namely, Wall Street firms and big money managers.

But there is something much deeper involved. Social Security is based on a principle that is considered subversive and that has to be driven out of people's heads: the principle that you care about other people. Social Security is based on the assumption that we care about each other, that we have a communal responsibility to take care of people who can't take care of themselves, whether they're children or the elderly. We have a social responsibility to pay for schools, to ensure day care, and to guarantee that whoever is taking care of children—including mothers—will be supported for doing so. That's a community responsibility and, in fact, the community benefits from it collectively. Maybe each individual can't say, "I benefit from that kid going to school," but as a society we benefit from it. And the same is true of caring for the elderly. But that idea has to be driven out of people's heads. There is huge pressure to turn people into pathological monsters who care only about themselves, who don't have anything to do with anyone else, and who therefore can be very easily ruled and controlled. That's what lies behind the attack on Social Security. And it reflects a deep imperative that runs through the whole doctrinal system.

Social Security was created in response to pressure from popular, organized social movements—the labor movement and others—that were based on the idea of solidarity and mutual aid. If you go back to Adam Smith, whom we're supposed to revere but not read, he assumed that sympathy was the core human value, and society should therefore be constructed so that this natural human dedication to sympathy and mutual support will be satisfied. In fact, his main argument for markets was that they would, under conditions of perfect liberty, lead to perfect equality. In fact, Smith's famous phrase "the invisible hand," which everyone totally misuses, appears only once in *The Wealth of Nations*, in the context of an argument against what we now call neoliberalism.[5] He says that if English manufacturers and investors imported from abroad and invested overseas, rather than here, it would be harmful to England. In other words, if they followed what are now called the principles of Adam Smith, it would be harmful to England. He said, however, there was no reason to worry about that because "upon equal or nearly equal profits, every wholesale merchant naturally prefers the home-trade to the foreign trade of consumption." That is, British capitalists will individually prefer to use domestically produced goods and to invest at home. So, therefore, as if "led by an invisible hand to promote an end which was no part of his intention," the threat of what's now called neoliberalism will be avoided. The economist David Ricardo made a rather similar argument. Smith and Ricardo both realized that none of their theories would work if you had free capital movement and investment.[6]

At one time, the principle of solidarity was taken for granted. It was a fundamental feature of popular movements. You're working for each other. That's why "Solidarity Forever" is a working-class slogan. And ever since the 1930s, the privileged and wealthy have been dedicated to trying to eliminate this principle. You have to destroy unions, you have to destroy interaction among people, you have to atomize people so they don't care about each other. And that's what really lies behind the attack on Social Security.

How do you deconstruct the idea that the United States is "bringing democracy" to Iraq?

It takes a minute's thought to see that there is no possible way that the United States and Britain would permit a sovereign, democratic Iraq. Just think what policies a democratic Iraq would follow. First, the state would have a Shiite majority, so it would probably shore up relations with Iran, which also has a Shiite majority. There is also a very substantial Shiite population in Saudi Arabia in the regions where the oil fields are located. A Shiite-dominated independence in Iraq, right next door, is very likely to elicit reactions in the Shiite regions of Saudi Arabia, which could very well mean that the core of the world's energy resources will be under the control or influence of an independent Shiite government. Is the United States going to allow that? It's unimaginable.

Second, an independent Iraq would try to recover its historic place as a leading force, maybe the leading force, in the Arab world. What is that going to mean? Iraq will rearm and will probably develop weapons of mass destruction, first as a deterrent and, second, to counter the main regional enemy, Israel. Is the United States going to sit by and allow that? The chances that the United States and its British attack dog will sit by quietly and allow any of these things to happen are so remote that you can't even discuss it. U.S. and British planners can't possibly be conceiving of a democratic Iraq. It's inconceivable.

In your writings and talks you quote from the New York Times, *the BBC, and other mainstream media. Critics of your position would say, "On the one hand, he's saying that the media are heavily biased in favor of existing power institutions and elites. On the other hand, he's getting his facts from those very media."*

I use them all the time. If I could read only one newspaper, it would be the *New York Times*. The *Times* has more resources and more coverage than any other newspaper, as well as some perfectly good correspondents. But that doesn't change anything. The major media do report information; they must, for a number of reasons. One is that their primary constituency requires it. Their primary constituency consists of economic managers, political managers, and doctrinal managers—the educated class, the political class, those who run the economic system. These people need a realistic picture of the world. They own it, they control it, they dominate it, they have to make decisions in it, so they'd better understand something about it. That's why, in my opinion, the business press tends to have better reporting than the other national press. Quite often you find stories in the *Wall Street Journal* or the *Financial Times* going into considerable depth in exposing corruption—not just robbery but the way the system undermines fundamental human needs. You are much more likely to read these stories in the *Wall Street Journal* than in the so-called liberal press, because that constituency has to have a reasonably realistic conception of the world. There is a doctrinal slant to what's reported to make sure readers see the facts in the right way, but the basic facts are there.

Furthermore, journalists generally have professional integrity. Typically they are honest, serious professionals who want to do their job properly. None of that changes the fact that most of them reflexively perceive the world through a particular prism that happens to be supportive of concentrated power.

One of our most cherished beliefs is that we have a free press. How free is the free press here?

The United States is, to my knowledge, unique in its guarantees of freedom of the press. The government in the United States has fewer options and less ability to control the press than in any other country I know. In England, for example, the government can raid the offices of the BBC and take its files. It can't do that in the United States. The government can't send the police into the offices of the *New York Times.* In England last year, the government investigated the BBC because it claimed that a reporter had gone too far in criticizing a completely deceitful government dossier on Iraq.[7] The reporter said that evidence of Iraqi weapons of mass destruction had been "sexed up." There was a huge uproar. Then a government-led review, the Hutton Report, came out, condemning the BBC and exonerating the government, and there was a huge public outcry about that, too. But that's the wrong focus. The outcry should have been over the fact that there was an inquiry at all. What right does the government have to carry out an inquiry into whether the media are reporting the facts the way it wants them to be reported? The very fact that the inquiry took place is a function of the very low commitment to freedom of speech in England.

The BBC, though, is regulated by the state and has a license issued by the state.

The radio airwaves are licensed in the United States, too, but that doesn't confer on the state any right to carry out official inquiries into whether they're doing their job in a way the government likes. The broadcast spectrum is owned by the public. But the fact that the government doesn't have much power to control the press doesn't mean that the press is free in practice; it tells you that it can be free if it chooses to be—though it may choose not to be. The press faces powerful pressures that induce it, and often almost compel it, to be anything but free. After all, the mainstream media are part of the corporate sector that dominates the economy and social life. And they rely on corporate advertising for their income. This isn't the same as state control but is nevertheless a system of corporate control very closely linked to the state.

In Necessary Illusions, *you say that citizens of democratic societies should "undertake a course of intellectual self-defense to protect themselves from manipulation and control."[8] Could you give some examples of what people might do?*

Intellectual self-defense is just training yourself to ask the obvious questions. Sometimes the answers will be immediately apparent; sometimes it will take a little work to find them. When you read that 100 percent of commentary agrees on something, whatever it is, you should immediately be skeptical. Nothing is that certain, even in nuclear physics. So if all the commentators say that the president's goals in Iraq are to bring democracy to the benighted citizens of a sovereign Iraq, and only differ on whether these noble and inspiring goals can be achieved, you should take the five minutes of reflection required to see that this can't possibly be true. And if 100 percent of educated opinion takes for granted something that cannot possibly be true, what does that tell you about the core doctrinal and cultural institutions? It tells you quite a lot.

You don't have to go back to David Hume to understand this, but he rightly observed that "*force* is always on the side of the governed, the governors have nothing to support them but opinion. It is, therefore, on opinion only that government is founded; and this maxim extends to the most despotic and most military governments, as well as to the most free and most popular."[9] In other words, in any state, whether a democratic state or a totalitarian state, the rulers rely on consent. They have to make sure that the people they are ruling do not understand that they actually have the power. That is the fundamental principle of government. Governments have all sorts of means to control the governed. In the United States, we don't use the stake, club, or torture chamber; we have other means. Again, it doesn't take special skills to figure out what they are, and that's all part of intellectual self-defense.

Let me give you another example. The *Washington Post* has a section called *KidsPost*. It's news of the day for children. Somebody sent me a clipping from *KidsPost* right after the death of Yasir Arafat. And it said in simple words pretty much what the main articles were saying in complicated words, but it added something that the complicated articles would know they couldn't get away with. It said, "[Arafat] was a controversial man, beloved by his own people as the symbol of their fight for independence. But to create a Palestinian homeland he needed land that is now part of Israel. He carried out attacks against the Israeli people that made many people hate him."[10] What does that mean? That means the *Washington Post* is telling children that the Occupied Territories are part of Israel. Even the U.S. government doesn't say that. Even Israel doesn't say that. But children are being indoctrinated into believing that the illegal Israeli military occupation is beyond question, because the territory they conquered is part of Israel. Intellectual self-defense should immediately have prompted a huge protest against the *Washington Post* for this disgraceful indoctrination of children. I don't read *KidsPost*, so I don't know if that goes on regularly, but I wouldn't be surprised.

What moves a citizen from being a passive onlooker, a spectator, to becoming engaged?

Take something recent in our history, the women's movement. If you had asked my grandmother if she was oppressed, she wouldn't have understood what you were talking about. If you had asked my mother, she knew she was oppressed and she was resentful, but couldn't openly question it. She wouldn't allow my father and me to go into the kitchen because that wasn't our job; we were supposed to be doing important things like studying, while she did all the work. Now ask my daughters if they're oppressed; there's no discussion. They'll just kick you out of the house. That's a significant change that's taken place recently, a dramatic change in consciousness and in social practice.

Walk down the halls of MIT today. Forty years ago you would have seen only well-dressed white males who were respectful to their elders, and so on. You walk down the halls today, half the people you see are women, a third are minorities, people are casually dressed. Those are not insignificant changes. And they have occurred throughout the society.

Are the hierarchies breaking down?

Of course. If women don't have to live like my grand-mother or my mother, hierarchies have broken down. For example, I learned recently that in the town where I live in Massachusetts—a professional, middle-class town, with lawyers, doctors, and so on—the police department has a special section that does nothing but answer 911 calls related to domestic abuse. Did anything like that exist thirty years ago, or even twenty years ago? It was inconceivable. It was none of anybody's business if somebody wanted to beat up his wife. Is that a change in hierarchy? Absolutely. Furthermore, it's only one part of a very broad set of social changes.

How does the change take place? Just ask yourself, how did the change take place from my grandmother to my mother to my daughters? Not through some benevolent ruler who passed laws granting rights to women. A lot of it was sparked by the young activist movements of the left. Take a look at draft resistance movement in the 1960s. Draft resisters were doing something very courageous. It's not easy for an eighteen-year-old kid to decide that he's going to risk losing his promising career and possibly spend years in jail or flee the country and possibly never be able to come back. That takes a lot of guts.

Well, it turns out that the youth movements of the 1960s, like the broader culture, were extremely sexist. You may remember the slogan, "Girls don't say no to boys who won't go," which was on posters at the time. Young women who were part of the movement recognized there was something wrong with the fact that women were doing all the office work and so on, while the men were going around parading about how brave they were. They began to regard the young men as oppressors. And this was one of the main sources of the modern feminist movement, which really blossomed at the time.

At some point, people recognize what the structure of power and domination is and commit to doing something about it. That's the way every change in history has taken place. How that happens, I can't say. But we all have the power to do it.

How do you know your mother felt oppressed? Did she ever say so?

Clearly enough. She came from a poor family with seven surviving children—a lot of children didn't survive in those days. The first six surviving children were girls. The seventh was a boy. The one boy went to college, not the six girls. My mother was a smart woman, but she was only allowed to go to normal school, not to college. And she was surrounded by all these guys with Ph.D.'s, my father's friends, and she very much resented it. For one thing, she knew that she was much smarter than they were. In fact, when I was a kid, whenever there was a party, the men would go in the living room, the women would sit around the dining room table and have their own conversations. As a kid I always drifted to the women's place, because they were talking about interesting things. They were lively, interesting, intelligent, political. The men, who were all Ph.D.'s, big professors and rabbis, were talking nonsense mostly. My mother knew it and she resented it, but she didn't think there was anything that could be done about it.

Thinking about protest movements, as I travel across the country, I often hear people say, "People in the United States are too comfortable. They have it too easy. Things will have to get much worse before there is protest."

I don't think that's true. Serious movements sometimes come from people who really are oppressed and other times it comes from sectors of privilege. We just spoke about the resistance movement. The kids involved with that were privileged college students, almost all of them from elite schools. But within those sectors of privilege, a spark was lit and these kids played a big role in changing the country. They infuriated the rich and the powerful. Take a look at the newspapers then. They're full of all sorts of hysterical screeching about bra burning and all these horrible things that were going on, undermining the foundations of civilization. But, in reality, the country was becoming civilized.

Take a look at SNCC, the Student Nonviolent Coordinating Committee, which was at the leading edge of the civil rights movement—the people who were really on the line, not the ones who showed up for a demonstration now and then but the ones out there every day, sitting at lunch counters, traveling on freedom buses, getting beaten up or in some cases killed. For the most part, the students in SNCC came from the elite colleges, like the college where Howard Zinn was teaching, Spelman, and where he was kicked out because he supported the students in their efforts.[11] Spelman was a black college, but an elite black college. Obviously not all the students in the movement came from privileged backgrounds, but they were certainly a leading part of this struggle.

And the same is true if you look at other movements. It's a mixture of privileged and oppressed coming to consciousness. Take the women's movement again. A lot of it began with consciousness-raising groups, women talking to each other and saying, "Look, life doesn't have to be like that." That was an early part of it, and it's a necessary part of any social movement. On the part of the oppressed, it's necessary to recognize that oppression is not just unpleasant but also wrong. And that's not so simple. Established practices and conventions are usually taken for granted, not questioned.

To recognize that there is nothing necessarily legitimate about power is a big step no matter which side of the equation you are on. A recognition that you are beating someone can be very enlightening. For those holding the club, it's a big step to say, "Look, there is something wrong with the fact we're holding the club." That recognition is the beginning of civilization. If the *New York Times* and its educated readers ever get to the stage that they think there is something wrong about carrying out the vicious war crimes that the *Times* is depicting on the front page, that's when the educated classes will begin to become civilized.

In your appearance with William F. Buckley on Firing Line *in 1969, you talked about guilt. You said, "I'm not interested in simply throwing blame around and giving marks. I think the beginning of wisdom in this case"—you were talking about Vietnam—"is recognizing what we stand for in the world, what we're doing in the world. And I think when we do recognize that, we will feel an enormous sense of guilt. One should be very careful not to let confessions of guilt overcome the possibility of action."*

I think it's an experience we've all had. You say, "Oh, yes, I did something terrible, I lament it. I'm not going to do anything about it. I've now expressed my guilt. The end." That happens all the time. But the crime isn't over when you express your guilt. You did something wrong, it had consequences. What are you going to do about it? Guilt can be a way of preventing action. You comfort yourself by saying, "Look how noble I am. I confessed that I did something wrong, and now I'm free."

You find this kind of thinking all the time. Take the case of Iraq. Right now, the United States is essentially coercing other countries into forgiving Iraq's debt.[12] That's the right thing to do. Everyone should forgive Iraq's debt, because it's what's known as "odious debt." Odious debt is debt that is forced on people under a system of coercion. For example, if the corrupt generals who run some society run up an enormous debt, is it the duty of the people of the country to pay it off? No. That's odious debt and should be eliminated.

The concept of odious debt was invented when the United States conquered Cuba—which historians here call the liberation of Cuba, meaning the conquest of Cuba to prevent them from liberating themselves. After taking over Cuba, the United States didn't want to pay Cuba's debt to Spain and they correctly pointed out that it was odious debt, incurred by Cuba under coercive conditions. The same thing happened in the Philippines. Of course, the real motivation was to absolve the United States from having to pay the debt of the countries they had just taken over. The same thing is happening now in Iraq. The United States has taken over Iraq, and doesn't want to have to pay the debt.

In reality, the United States should be paying huge reparations to Iraq. So should Britain, so should Germany, so should France, so should Russia, and all the other states that supported Saddam Hussein. These countries have tortured Iraq for a long time, in fact back to the time when Iraq was created by the British in the early 1920s. John F. Kennedy apparently sponsored a military coup in 1963 that put Saddam Hussein's Baathist party in power.[13] Since then, the U.S. record with regard to Iraq has been horrendous. The State Department keeps a list of states that sponsor terrorism. Only one country has ever been taken off the list—Iraq in 1982—because the Reagan administration, basically the guys in office again now under Bush II, wanted to be able to supply Saddam Hussein with weapons and aid "without Congressional scrutiny."[14] So Iraq was suddenly a state that didn't sponsor terrorism, and the United States could provide aid for agribusiness exports, for developing weapons of mass destruction, and all sorts of wonderful things.

After Hussein's atrocities against the Kurds, against Iran, against Iraqis—which we now denounce—the United States continued to support Saddam Hussein. After the 1991 Gulf War, when a Shiite rebellion broke out, Bush I allowed Saddam Hussein to crush it. So when Thomas Friedman of the *New York Times* now writes columns about how, gosh, he discovered these mass graves in Iraq and feels terrible, he should acknowledge that he knew all about the graves at the time and that the U.S. government was complicit.[15] And then came more than ten years of sanctions, which killed more people than Saddam Hussein ever did, and devastated the society.[16] And then came the invasion, which has led to the deaths of maybe a hundred thousand people.[17]

Put it all together. We owe Iraq huge reparations. Getting rid of the odious debt is okay, but that's for our benefit. Paying reparations is not.

The same thing applies to Haiti, the poorest country in the hemisphere, which is almost at the verge of extinction. Who's responsible for that? The two main criminals are France and the United States. They owe Haiti enormous reparations because of actions going back hundreds of years. If we could ever get to the stage where somebody could say, "We're sorry we did it," that would be nice. But if that just assuages guilt, it's just another crime. To become minimally civilized, we would have to say, "We carried out and benefited from vicious crimes. A large part of the wealth of France comes from the crimes we committed against Haiti, and the United States gained as well. Therefore we are going to pay reparations to the Haitian people." Then you will see the beginnings of civilization.

Let's go back to oppression for a minute. Say you're abusive toward me. I experience it firsthand. Isn't it much more difficult to understand imperialism because that's happening somewhere out there, far away, and I don't know much about it?

Not only that but the logic is reversed, so that people here feel they're the ones who are oppressed. The line of the soldiers who carried out atrocities in Iraq is that the Iraqis did it to us, so we're going to do it to them. What did the Iraqis do to us? 9/11. Of course, the Iraqis had nothing to do with it, but the feeling still is that we're the ones under attack; they're the ones who are attacking us. And that inversion goes on all the time.

Take Ronald Reagan and his rhetoric about "welfare queens." We poor people, like Reagan, are being oppressed by these rich black women who drive up in Cadillacs to get their welfare checks. We're being oppressed. And in fact that's a strain that goes right through U.S. history. There's a book by Bruce Franklin, a literary theorist, that traces this strain through American popular literature, going back to the colonists. We are always just on the verge of extinction. We're being attacked by demonic enemies who are just about to overwhelm us, and then, at the last minute some superhero or amazing weapon appears and we're able to save ourselves.[18] But, as Franklin points out, it's consistently the case that the people who are about to exterminate us are the ones who are under our boot. We've got our boot on their necks, and that means they're about to exterminate us.

Like the "merciless Indian savages," as Native Americans were described in the Declaration of Independence.

Exactly. "Merciless Indian savages" are about to exterminate us. Then it was the blacks. Then it was the Chinese immigrants. Jack London, a progressive writer, a leading socialist figure, wrote stories in which he literally called for exterminating the entire population of China by bacteriological warfare because that's the only way we can save ourselves. They are sending over these people who we think are coolies building the railroads and laundrymen washing our clothes, but it's all part of a plan to infiltrate our society. There are hundreds of millions of them, and they're going to destroy us. So we have to defend ourselves, and the only way we can do it is by totally exterminating the Chinese race through bacteriological warfare.

Or take Lyndon Johnson. Johnson, whatever you think about him, was a kind of populist. He was not a fake Texan like George Bush but a real one. And he said, "Without superior air power America is a bound and throttled giant; impotent and easy prey to any yellow dwarf with a pocket knife."[19] In one of his main speeches to U.S. troops in Vietnam, Johnson said plaintively,

"There are three billion people in the world and we have only two hundred million of them. We are outnumbered fifteen to one. If might did make right they would sweep over the United States and take what we have. We have what they want."[20] That is a constant refrain of imperialism. You have your jackboot on someone's neck and they're about to destroy you.

The same is true with any form of oppression. And it's psychologically understandable. If you're crushing and destroying someone, you have to have a reason for it, and it can't be, I'm a murderous monster. It has to be self-defense. I'm protecting myself against them. Look what they're doing to me. Oppression gets psychologically inverted: the oppressor is the victim who is defending himself.

I was just thinking, we've been doing interviews for twenty years now. Do you ever feel like Sisyphus of Greek legend, rolling the boulder up the hill, and just having it roll back down?

Not really. For one thing, almost all of us are so privileged and so free that to feel that there is anything difficult about our lives is outrageous. Whatever repression or vituperation we have to confront is nothing compared to what people face anywhere else. It's a kind of luxury that we should never grant ourselves. But that aside, there have been changes. So you're rolling the rock up the hill but making progress, too.

You're sometimes like Cassandra, constantly issuing warnings. Your latest book, Hegemony or Survival, *starts and ends on pretty dire notes.*[21]

I think the warnings are realistic. I start off *Hegemony or Survival* by quoting Ernst Mayr, probably the world's most distinguished biologist, and end by quoting Bertrand Russell, the most distinguished philosopher of the twentieth century, and their points are accurate. You can easily add others. *Dædalus,* the journal of the American Academy of Arts and Sciences, the peak of establishment respectability, recently had an article by two highly respected mainstream strategic analysts, John Steinbruner and Nancy Gallagher, on what's called the transformation of the military, which includes the militarization of space.[22] The militarization of space means, in effect, placing the entire world at risk of instant annihilation with no warning. What do Steinbruner and Gallagher suggest as a remedy? They hope that a coalition of peace-loving states led by China will coalesce to counter U.S.

militarism and aggressiveness. That's the only hope they see for the future. One of the interesting aspects of this argument is the despair or contempt—I don't know what the right word is—for U.S. democracy: the United States can't be changed internally, so let's hope China will rescue us. It is unprecedented to hear this kind of thinking at the heart of the establishment. What I wrote in *Hegemony or Survival* is mild in comparison.

DEMOCRACY AND EDUCATION

John Dewey, one of the leading thinkers of the twentieth century, had a strong influence on you in your formative years. Your parents sent you to a Deweyite school in Philadelphia.

My father ran the Hebrew school system in Philadelphia, where I lived, and it was run on Deweyite lines, which meant trying to focus on individual creativity, joint activities, stimulating projects. I taught there, as well. In the school I attended, we covered all the regular subjects, but with an emphasis on the child's concerns and commitments and creative engagement. There was no competition among students. I didn't even know that I was a so-called good student until I left the school to attend high school. In high school, everybody was ranked, so you found out where you were. It was just never an issue before.

What motivated your parents to send you to this school?

Partly it was because they worked, so I had to be at school all day. But really I wouldn't have wanted to be anywhere else. I started there at about eighteen months and went through eighth grade.

Tell me about your father. What was your relationship with him? And, of course, he was not only your first teacher but, it sounds like, your first employer also.

He was a Hebrew scholar. We had a very warm relationship. We didn't spend huge amounts of time together—during the days I was mostly at school or out in the street with friends—but the time we did spend together was significant and meaningful. On Friday nights we read traditional and modern Hebrew literature together. Since my parents were teachers and didn't work over the summer, we took long summer vacations. And my father would work during the day, but he would come out in the late afternoon, and we'd all go swimming together. By the time I was about, I guess, eleven or twelve, I started getting interested in his scholarly work. My father was just finishing up a Ph.D. dissertation on David Kimhi, the medieval Hebrew grammarian, which I remember reading. I also read his articles, and we would talk about them.

Do you think that helped to train your mind in some way in terms of mastering a complex language with a pretty dense grammar?

That's hard to say. It did get me interested in Semitic linguistics, which I studied in college. And it probably had some indirect influence on my getting into linguistics, but I can't really trace it.

In Propaganda and the Public Mind, *you said that "my intellectual achievement was retarded when I went to high school. I sort of sank into a black hole."*[1]

That's pretty accurate. Getting into high school was a bit of a shock. I went to an academic high school, very rigorous and disciplined. I disliked almost every aspect of it, aside from my friends. But I remember very little of it, whereas I remember elementary school and up to high school very vividly. I couldn't wait to get out.

After high school, I went to the local college in Philadelphia, the University of Pennsylvania. I had no thought of doing anything else except living at home, working, and commuting to school—and I was very much looking forward to it. The course catalog looked exciting and interesting. But I was disabused of that idea within a year or so. I found everything to be just a boring continuation of high school, and was pretty close to dropping out, in fact.

But at some point you met Zellig Harris, who was teaching linguistics at the University of Pennsylvania.

I actually met him through political contacts when I was about seventeen. I was a sophomore, toying with the idea of dropping out, and in fact spending very little time on academic work—I think I was probably majoring in handball by that time. I was very involved in the Zionist movement, specifically its binationalist, antistate wing. And it turned out Harris was a leading figure in that. He also happened to be a very charismatic and intellectually exciting person, whose other interests—anarchist thought, the anti-Bolshevik left, and so on—I was also trying to explore on my own.

I suspect, in retrospect, that Harris was trying to get me back into college. He didn't say it, but he suggested I take some of his graduate courses, and I did. There was a scattering of extremely good faculty in different fields: one in mathematics, one in philosophy, one somewhere else. By picking and choosing, you could get an exciting education without much formal structure. And Penn was loose enough so it didn't matter.

Did you ever actually get a piece of paper, a graduate degree?

I ended up getting all the degrees formally, but without having fulfilled the usual requirements. The linguistics department was fairly unstructured. Harris essentially ran it. In a way, it was advantageous to me that the school was not a very academically prestigious place, so that it didn't have heavy requirements and supervision and so on. You could kind of do what you wanted—at least I could.

So, including your early years, you've been a teacher for more than six decades. You've had thousands of students. What qualities do you look for in a student?

Independence of mind, enthusiasm, dedication to the field, and willingness to challenge and question and to explore new directions. There are plenty of people like that, but school tends to discourage those characteristics.

Does it ever happen that students are so in awe of you—I mean, you are well known—that they're reluctant to challenge some of your assertions?

Occasionally. It's sometimes true of the students who have come through traditional educations in Asian countries, for example. But at a place like MIT that's relatively rare. This is a science-based university, so students are, in fact, encouraged to pursue research, to challenge, to question.

As your career in linguistics developed, you also were becoming more involved politically. What did your parents think of that? Were they worried that you could get into trouble?

I had always been involved in politics, but by the 1960s they had to be worried, because I was getting arrested and facing jail and so on. When the issue of Israel and the Palestinians became central, especially after 1967, and there was just a huge flood of vilification, hatred, slander, denunciation, they were supportive of my views, but it was difficult for them. They lived in almost a Jewish ghetto, and they were upset at the hysterical slanders and personal attacks. My father even wrote responses in the Hebrew press to some of the charges and denunciations. It wasn't easy for them. In fact, I probably semiconsciously cut some corners as long as they were alive, just to spare them.

You were trained in the hard sciences, in which empirical evidence is paramount whereas ideology often doesn't require any evidence whatsoever.

In fact, real commitment to ideology denies and tries to avoid evidence. But I wasn't trained in the hard sciences. I have some background in the hard sciences— I even worked in mathematics for a while—but I don't want to exaggerate. As I said, I have almost no formal training in any field, including linguistics. I'm mostly self-educated. But I don't see any particular reason not to study history, society, and economics by essentially the same methods that one uses in the sciences. Empirical evidence is critically important. You're flooded with it. You have to try to select what's significant. You inevitably approach evidence with certain beliefs and principles, which you should keep open to question. The problems are different in history and in physics, but the method of approaching them ought to be about the same.

Sometimes you're described as an anarcho-syndicalist, and I've even heard you describe yourself as an old-fashioned conservative. How do you feel about those labels?

I don't use these labels but I do feel that my views grew out of the anarcho-syndicalist tradition. I think anarcho-syndicalism represents a reasonable approach to the general problems of human society. Of course you can't take anarchist doctrines and mechanically apply them. But workers' control of industry and popular control of communities seem to me to be a sensible basis for a complex society like ours. As for old-fashioned conservative, that term partly reflects my personal tastes in music and literature and so on, and partly my belief in the value of classical liberal doctrines. Again, they're not mechanically applicable to the modern world in the language in which they were formulated, but I think one should have a good deal of respect for Enlightenment ideals—rationality, critical analysis, freedom of speech, freedom of inquiry—and should try to amplify, modify, and adapt them to a modern society.

Lately we frequently hear that Enlightenment ideas are under attack, particularly in education, where abstinence is being taught rather than other forms of protective sex, creationism is being advocated, textbooks are being censored. Are you worried about this trend?

This is a very worrisome feature of U.S. culture. No other industrial country has anything like the degree of extremist religious beliefs and irrational commitments that you commonly find in the United States. The idea that you have to avoid teaching evolution or pretend you're not teaching it is unique in the industrial world. And the statistics are mind-boggling. Roughly half the population think the world was created a couple of thousand years ago. A huge percentage, maybe a quarter or so, say they've had a born-again experience. A substantial number of people believe in what's called "the rapture." Large majorities are convinced of miracles, the existence of the devil, and so on.

These strains go pretty far back in American history but in recent years they have come to affect social and political life to an unprecedented extent. For example, before Jimmy Carter, no U.S. president had to pretend to be a religious fanatic, but since then every one of them has. This has contributed to a genuine undermining of democracy since the 1970s. Carter, probably inadvertently, taught the lesson that you can mobilize a large constituency by presenting yourself, honestly or not, as a Bible-fearing, evangelical Christian. Up until that point, religious beliefs were people's personal concerns. There has been a conscious takeover of the electoral system by the public relations industry, which now sells candidates the way they sell commodities. And the image of a God-fearing, believing person of deep faith who is going to protect us from the threats of the modern world is one you can sell.

I work in radio, and we cannot play Allen Ginsberg's "Howl," arguably one of the great poems of the twentieth century, because it has a forbidden word. We can't play Bruce Cockburn's song "Call It Democracy," because he says something very unflattering about the IMF, or Bob Dylan's song "Hurricane," about the unfair imprisonment of Rubin "Hurricane" Carter, a famous boxer, which also uses a taboo word.

There's a big assault on freedom of speech everywhere, in radio, in universities. More than a dozen state legislatures are now considering legislation, which I suppose some of them will pass, to control what teachers and professors say in classrooms and to make sure teachers don't "indoctrinate students."[2] As one of the sponsors of the legislation explained, "80 percent or so of [professors] . . . are

Democrats, liberals or socialists or card-carrying Communists."[3] This is part of an old nativist strain that is now being converted into a weapon against whatever institutions are not completely bought or controlled. The universities are pretty right wing, but they're not wholly owned subsidiaries of the corporate sector, and that's unacceptable.

There has been a live and very important tradition of academic freedom in the United States, which shouldn't be denigrated. Academic freedom has been under assault, but it's been protected and defended. There were serious reversals in the early 1950s, but that was finally overcome and we have even seen some apologies and retractions on the part of the institutions for their past behavior. But academic freedom is constantly under assault. And now it's increasing as part of the effort to ensure ultra-right domination. Anything that's out of control has to be suppressed and disciplined.

Let me ask you about nuclear weapons. It was just announced that the United States is developing a new generation of them.

The signatories of the Treaty on the Non-proliferation of Nuclear Weapons (NPT) have an obligation to undertake good-faith efforts to eliminate nuclear weapons. That's part of the bargain by which other countries agreed not to develop nuclear weapons. All of the NPT countries have violated the agreement, but the recent steps by the Bush administration go far beyond mere nonadherence. These measures are being portrayed in an anodyne fashion: we're just going to improve the weapons and make them more secure. But in reality, we are probably moving toward resuming nuclear testing and developing more destructive weapons. This is especially dangerous given that the United States officially reserves the right to use nuclear weapons in a first strike, even against nonnuclear powers. We hear every day that nonnuclear countries might be moving toward going nuclear, and we certainly don't want that to happen. But for the nuclear states to violate the treaty is far more serious and dangerous. They've brought the world pretty close to destruction a number of times and are very likely to do so again.

The year 2005 marks the sixtieth anniversary of the atomic bombings of Hiroshima and Nagasaki. You were around sixteen at the time of the attacks. What was the effect on you?

I was a junior counselor in a Hebrew-speaking summer camp at the time, somewhere in the Poconos, near Philadelphia, where we lived. We just heard the news. And I remember very vividly being sort of doubly shattered by it: first of all, by the news and, second of all, by the fact that nobody cared, which just struck me as so amazingly unbelievable that I walked off into the woods and spent a couple of hours sitting there by myself thinking about it.

Was it perhaps because no one could conceptualize what it meant? It was just another big bomb?

I don't think so. It's not an unfamiliar phenomenon. Is it surprising that kids in a summer camp didn't pay much attention to the fact that there had been an atomic bombing? Let's go back a couple of months before that. In March 1945, there was an air raid on Tokyo, which was targeted because the Allies knew that they could easily destroy the city, which was largely made of wood. Nobody knows how many people were killed. Maybe one hundred thousand people were burned to death. Do you remember any discussion of that? In fact, the fiftieth anniversary of the firebombing passed with scarcely a mention.

As you look back over your many years of teaching and activism, what have you been trying to do?

My teaching and activism have different goals. In teaching and research, which are inseparable, my goal is to understand something about the nature of the human mind. I'm particularly interested in language, but as a kind of window into the nature of cognitive systems, systems of thought, interpretation, and planning. I have my own special interests. One of them is a topic that has been very hard to study until very recently, which is the extent to which characteristics of biological systems—and I take systems of thought and planning and language to be biological systems—can be determined by very general properties of physical law, mathematical principles, and so on. By now there are beginning to be insights into these questions. It's been pretty exciting work, for me at least, in the past few years.

As for activism, that's just elementary. There is an enormous amount of human suffering and misery, which can be alleviated and overcome. There is oppression that shouldn't exist. There is a struggle for freedom all the time. There are very serious dangers: the species may be heading toward extinction. I can't see how anybody can fail to have an interest in trying to help people become more engaged in thinking about these problems and doing something about them.

ANOTHER WORLD
IS POSSIBLE

CAMBRIDGE, MASSACHUSETTS (FEBRUARY 8, 2005)

We've talked about the surge of religious fundamentalism in this country. What do you think accounts for it?

It's not really a surge. This has been a deeply religious country for a long time. Actually, I hate to use the word *religious*. Part of the reason I don't like the word is that you could make the argument that organized religion is sacrilegious. It's based on very strange conceptions about the deity. If there were one, he wouldn't like it. But let's for the moment use the word. It's been a very religious country since its origins. New England was settled by extremist religious fundamentalists who regarded themselves as the children of Israel, following the orders of the war god whom they worshipped as they cleansed the land of the Amalekites. You read the descriptions of some of the massacres, like the Pequot massacre, and they are just like chapters out of the most genocidal parts of the Bible, which in fact the settlers quoted liberally. The Western expansion was driven by religious fundamentalism with pseudo-biblical origins. Spanish areas were conquered under the banner of destroying the heresy of papism.

Typically, there is an inverse correlation between extremist religious beliefs and industrialization: the greater the modernization, the less commitment there is to religious extremism. But in the United States, the correlation completely breaks down. It's like an underdeveloped society in this respect. I remember fifty years ago driving across the country, listening to the radio. I couldn't believe what I was hearing. Preachers raving, screaming—you just can't imagine something like that anywhere else.

As for changes in recent years, I don't think they have so much to do with the level of religious commitment as with the way in which religion has been brought into the political system and public life. We've talked about how every president since Carter has had to be "religious," but you can observe this process everywhere.

The teaching of evolution, which is just normal in every other country, is extremely difficult here. And it has been for a long time. I remember when my wife was in college in the late 1940s. She was taking a sociology course, and I remember her telling me that the instructor said, "The next section is going to be on evolution. You don't have to believe this, but you just ought to know what some people think." I doubt if that happens in any other industrial country. And this was not the Deep South. This was the University of Pennsylvania. So we can argue about the causes of religious extremism in the United States but it's an undeniable aspect of American exceptionalism, one of many.

One possible cause is that this has always been a very frightened country, as we discussed. There is an unusually strong sense of insecurity here, which might be related to the degree of religious fundamentalism. The United States is by far the most powerful and secure country in the world, but it's the one that feels most insecure. John Lewis Gaddis, the well-known historian, recently wrote a sympathetic account of Bush's national security strategy. He traces it back to early U.S. history, specifically to John Quincy Adams, who laid out the grand strategy for the conquest of the continent. The centerpiece of his argument is a famous letter that Adams wrote in 1818 justifying Andrew Jackson's conquest of Florida during the First Seminole War.[1]

Gaddis cites Adams's argument that it was necessary to attack the Florida area in order to protect American security because the area was a "failed state"—he actually uses the phrase—a kind of a power vacuum which threatened the United States.

But if you examine the actual scholarship, it's quite interesting. Gaddis certainly knows that the scholarly books he cites point out that Andrew Jackson's invasion of Florida had absolutely nothing to do with security. It was a matter of expansion, a bid to take over the Spanish settlements. And the only threats were "lawless" Indians and runaway slaves. The Indians were lawless because they were being driven out of their homes and murdered, and the slaves were running away because they didn't want to be slaves. There were cases of Indian attacks on American settlements, but these were in retaliation against American attacks. That was called terror, of course, and we had to defend ourselves against it by conquering Florida.

Gaddis's point is that a guiding principle of U.S. history is that the only way to gain security is through expansion. Since we hadn't expanded into Florida, we were insecure, and the way to gain security was to expand. The fight to gain Florida turned out to be a real war of extermination—a vicious, brutal, murderous war. But that's fine, because we were doing it for security. You can trace that theme right up to the present. The same arguments are being made right now for the militarization of space: the only way we can have security is by expanding into and ultimately owning space.

Another aspect of religion in the United States is dissent and opposition, which was reflected in the Central American solidarity movement of the 1980s and during the recent invasion of Iraq when some clergy and churches spoke out.

Central America was a very striking case, because the United States was basically at war with the Catholic Church. The Catholic Church in Latin America in the 1960s and 1970s had really shifted its traditional vocation. It had adopted aspects of liberation theology, and had recognized what's called "the preferential option for the poor." Priests, nuns, and lay workers were organizing peasants into communities, where they would read the Gospels and draw lessons about organization that they could use to try to take control of their own lives. And, of course, that immediately made them bitter enemies of the United States, and Washington launched a war to destroy them. For example, one of the publicity points of the School of the Americas, which changed its name in 2000 to the Western Hemisphere Institute for Security Cooperation, is that the U.S. army helped "defeat liberation theology," which is accurate.[2]

The Central American solidarity movement in the United States in the 1980s was something totally new. I don't think there has been anything like this in the history of Europe. I don't know of anyone in France who went to live in an Algerian village to help people and protect them against marauding French paratroopers, but tens of thousands of Americans went down in the 1980s and protected people under assault from the United States. The center of the U.S. solidarity movements in the 1980s was not in the elite universities but in the churches, including churches in the Midwest and in rural areas. It wasn't like the 1960s. It was quite mainstream.

It's interesting to look back at what was happening at the time. Here's this supposedly very religious country, the United States, going to war against organized religion. And the reason was that the church was working for the poor. As long as religion is working for the rich, it's fine; but not for the poor.

Let's shift gears, and talk about the economics of empire. The U.S. dollar is weak now, government deficits are up, individual consumer debt is up, credit-card interest rates are rising, personal savings rates are at all-time lows, and foreign investors are financing the U.S. debt by buying Treasury securities. How long can this be sustained?

We don't really know. Actually, the debt situation is complicated. Household debt is out of sight, but corporate debt is very low. In fact, corporations are making huge profits. That's part of the shift in the way economic planning is carried out, to benefit the superrich and the corporations and to harm the general population. In fact, the ratio of taxed income to gross domestic product is close to an all-time low, and it's skewed toward the general public, much more so than before. Corporations barely pay taxes. The corporate tax rate is already very low, but corporations have worked out an array of complicated techniques so they often don't have to pay taxes at all.

To give you an example, in the mid-1990s there was a lot of excitement about so-called emerging markets in Latin America. Out of curiosity, I started to read U.S. Department of Commerce reports about foreign direct investment (FDI) in Latin America. It turns out that foreign direct investment in Latin America did surge during the mid-1990s, but the composition of it was extremely interesting. Consistently about 25 percent of FDI was going to Bermuda, around 15 percent was going to the British Cayman Islands, and about 10 percent to Panama. That's roughly 50 percent of what they're calling foreign direct investment, and it certainly was not going to build steel plants. This was just money flowing into various tax havens. Most of the rest was going for mergers and acquisitions and so on. These are huge sums. The scale of sheer robbery by corporate power is enormous.

In any event, corporations and rich people barely pay taxes, so they're doing fine. But the general population has gone through thirty years of either stagnation or decline in real wages, with people working longer hours with fewer benefits. I don't think there's been a period like this in American history.

The United States is still a very rich country. It's got enormous advantages, of scale, resources, anything you can think of. But it's being subjected to domestic policies that are frightening. Conservative economists are tearing their hair out watching the Bush administration purposely drive the country into incredible debt. The idea of the Bush administration is to transfer costs to future generations. That's basically their plan. Their values are to serve the rich and the powerful, and to transfer the costs to the general population in the future generations. When you talk about their "moral values," that's what they are.

Take, say, health-care costs, which are going out of sight. The United States has a highly inefficient health care system, the worst in the industrial world, with huge expenses, much higher than in any other country, and with relatively poor outcomes. The costs are rising even further, partly because of the tremendous power of pharmaceutical corporations, and partly because of all the administrative costs of a privatized health care system. This is a real crisis, unlike the Social Security crisis, which doesn't exist.

Why are they going after Social Security but not the medical system? I think it's straightforward. Take somebody like me, an overly well-paid college professor, now retired. I receive Social Security, but it doesn't amount to that much of my income. I get fantastic medical care, because I'm rich and medical care is rationed by wealth. If you're rich, the system is working just right. The insurance companies, the health maintenance organizations, the pharmaceutical corporations are doing just great. Wealthy people are doing fine. If most of the population can't get decent medical care, that's not our problem. If health care costs are astronomical, too bad.

The administration recently announced that they're

going to cut back federal funding for Medicaid.[3] But that only harms poor people, so it's fine. Social Security, though, that's a real problem because it does nothing for the rich. It's a useless system.

As for how long this can all go on, I don't think anybody really knows. There could be a revolt, there could be a huge economic crash, there could be adventurism leading to a major war.

Speaking of health care, you told me recently about a visit you had to the clinic here at MIT.

I've been at MIT for a long time, so both my wife and I know a lot of the medical staff. They say they're now spending maybe 40 percent of their time filling out forms. They're under constant supervision and control. They're wasting huge amounts of time doing tons of paperwork that isn't necessary. And those are all costs.

Economists have highly ideological ways of measuring costs. I'm sure you've had this experience, but suppose you want to order an airline ticket, fix a mistake on your bank statement, suspend your newspaper delivery, or whatever it may be. It used to be you could make one call, talk to somebody, and take care of the problem in two minutes. Now what happens is you call a number, and you get a recorded message that says, "Thank you for calling. We appreciate your business. All of our agents are busy." First of all, you get a menu that you can't understand, and it doesn't have what you want on it anyway. So then it says wait for somebody. Then you wait and they play a little song, and every once in a while this recorded voice comes on asking you to keep waiting—and you can sit there for an hour waiting. Finally somebody comes on, who is probably in India, doesn't know exactly what you're talking about, and then maybe you will get what you want, but maybe not.

The way economists measure this, it's highly efficient. It increases productivity, and productivity is what's really important, because that's what makes life better for everyone. Why is it efficient? Because businesses are saving money. The costs are being transferred to consumers, of course, but that's not measured. Nobody measures the amount of time that it takes you to get some simple task done or to correct errors, and so on. That's just not counted. If we were to count such real costs, the economy would be extremely inefficient. But the ideological principle is that you count only the costs that matter to rich people and corporations.

A recent study by the Harvard Medical School and Public Citizen compared the U.S. and Canadian health care systems.[4] The study found that the United States is spending several hundred billion dollars a year in excess administrative costs. One of the things they did was to compare one of the main hospitals in Boston with a leading hospital in Toronto. When the research team visited the Toronto hospital, they wanted to examine the billing department. Nobody knew where it was. Finally they found a little office down in the basement somewhere that had a billing department for U.S. citizens who were coming to Canada. In Boston, the billing office takes up a whole floor full of accountants, computers, and paperwork. All of that adds up.

You said in a talk at the Harvard Trade Union Program that the United States does have a form of universal health care. They're called emergency rooms. Could you explain that?

Most states have laws stipulating that if you go to an emergency room, they have to take care of you, even if you don't have health insurance. So that's universal health care. Sometimes the emergency rooms are overflowing and you can't get in. Or if you do get in, you may have to wait a long time before any doctor can help you. The father of a friend of mine was very sick, and he had to take him to the hospital. The father didn't have health insurance, and this friend actually sat there for three days bringing him food and taking care of him *before* the doctors saw him. His father wasn't dying, he just needed care.

A couple of months ago, I had uncontrollable nose-bleeds. They weren't life threatening but they were a damned nuisance. I called MIT, and they told me to go over to Lahey Clinic, which is a very fancy hospital complex for very elegant types near where I live. So I went over to the Lahey emergency room, and I sat there for a couple of hours. Finally, I was treated by a specialist, who was far more skilled than anybody I needed. The emergency care system is not giving people the kind of care they need. It wastes an enormous amount of time. It's not preventive care, figuring out how to avoid getting sick in the first place. It's the most expensive, most inefficient kind of universal health care system you can imagine.

Downtown Boston has two big hospitals right next door to each other—Boston City Hospital, run by the city, and a private hospital that is part of the Tufts medical system. I was talking to the staff at Boston City Hospital a while back, and I was told that if an ambulance drives up to the Tufts Medical Center, often it will be sent over to the city hospital. The reason is, if an ambulance brings a sick patient into the hospital, the hospital has to take care of him. And if the patient is indigent, the hospital is going to have to pay for it. They'd rather have the city pay, so they send them next door.

This seems like an enormous wedge issue in terms of organizing popular support. Forty-five million Americans have no coverage whatsoever, yet people seem more worried about Janet Jackson's breast being revealed at the Super Bowl.

I don't know that they're more worried about that. I think people are very concerned about health care. Whenever the question is asked in polls, it turns out that people rank it as a very high concern. I think about three-fourths of the population, the last I noticed, wants higher health care expenditures.[5]

I know about those polls, but I'm struck that hundreds of thousands of people turned out to protest the Iraq war. Yet health care, which affects everybody, doesn't seem as urgent an issue.

The hundreds of thousands in the streets is a one-shot affair. You organize a demonstration and people come out. Then most of them go back home, and continue with their lives. Health care is a different issue. You can't get it with one demonstration. You have to have a functioning democratic society, with popular associations, unions, and political groups working on it all the time. That's the way you organize people to get health care. But that's what's lacking.

The United States is basically what's called a "failed state." It has formal democratic institutions, but they barely function. So it doesn't matter that approximately three fourths of the population think we ought to have some kind of government-funded health care system. It doesn't even matter if a large majority regards health care as a moral value. When commentators rave about moral values, they're talking about banning gay marriage, not the idea that everyone should have decent health care. And the reason is that it's not in their interest. They're like me; they get fine health care. What do they care? For the large majority of the population, though, lack of health care is a major issue, and it's becoming an even more serious one. When Medicaid is destroyed, as it probably will be, that's going to really harm people. But those people are unorganized. They're not in unions, they're not in political associations, they don't participate in any political parties. The genius of American politics has been to marginalize and isolate people. In fact, one of the main reasons behind the passionate effort to destroy unions is that they are one of the few mechanisms by which ordinary people can get together and compensate for the concentration of capital and power. That's why the United States has a very violent labor history, with repeated efforts to destroy unions anytime they make any progress.

In fact, Missouri and Indiana have recently abolished the right for public-sector workers to engage in collective bargaining.[6]

The federal government has done pretty much the same. Part of the Bush administration's Department of Homeland Security scam was to strip a hundred and eighty thousand government workers of union rights.[7] Why? Are they going to work less efficiently if they're unionized? No. It's just that you have to eliminate the threat that people might get together and try to achieve things like decent health care, decent wages, or anything that benefits the population and doesn't benefit the rich. You can almost predict policy by that simple principle: Does it help rich people or does it help the general population? And from that you can virtually deduce what's going to happen next.

You are often asked about possibilities for the future. One source of hope in the world today for some people is the World

Social Forum, a gathering of tens of thousands of activists from around the world each year. The theme of the forum is "Another world is possible." I'm interested in this formulation. It's not a question but an affirmation. What might another world look like that you would find attractive?

You can start with small things. For example, I think it would be an improvement if the United States became as democratic as Brazil. That doesn't sound like a utopian goal, does it? But just compare the two most recent elections here and in Brazil. In Brazil, where there are vibrant popular movements, people were able to elect a president, Lula, from their own ranks. Maybe they don't like everything Lula's doing, but he's an impressive figure, a former steelworker. I don't think he ever went to college. And they were able to elect him president. That's inconceivable in the United States. Here you vote for one or another rich boy from Yale. That's because we don't have popular organizations, and they do.

Or take Haiti. Haiti is considered a "failed state," but in 1990 Haiti had a democratic election of the kind we can only dream of. It's an extremely poor country, and people in the hills and the slums actually got together and elected their own candidate. And the election just shocked the daylights out of everyone, which is why in

1991, there was a military coup, supported by the United States, to crush the democratic government. For us to become as democratic as Haiti doesn't sound very utopian. For us to have a medical care system like Canada's is not reaching for the stars. For us to have a society in which the wealth of the country isn't concentrated in the hands of a tiny elite isn't utopian.

And you can go on from there to much more far-reaching goals. Many of the basic institutions of our society are totally illegitimate. Do corporations have to be controlled by management and owners and dedicated to the welfare of shareholders instead of being controlled by the people who work in them and dedicated to the community and the workers? It's not a law of nature.

NOTES

1. IMPERIAL AMBITIONS

1. White House, *The National Security Strategy of the United States of America,* released 17 September 2002, online at http://www.whitehouse.gov/nsc/nss.html.

2. Linda Feldmann, *Christian Science Monitor,* 14 March 2003.

3. Peter Ford, *Christian Science Monitor,* 11 September 2002. See also polls cited in Noam Chomsky, *Hegemony or Survival* (Owl Books, 2004).

4. Noam Chomsky, "Confronting the Empire," 2 February 2003, online at http://www.chomsky.info/talks/20030201.htm.

5. Dean Acheson, *Proceedings of the American Society of International Law,* no. 13/14 (1963).

6. *Foreign Relations of the United States* (1945), vol. 8, p. 45.

7. Andy Webb-Vidal, *Financial Times* (London), 14 January 2005.

8. Stephen Farrell, Robert Thomson, and Danielle Haas, *The Times* (London), 5 November 2002.

9. Robert Olsen, *Middle East Policy* 9, no. 2 (June 2002).

10. Richard Wilson, *Nature* 302, no. 31 (March 1983).

11. Imad Khadduri, "Uncritical Mass," (manuscript, 2003). Michael Jansen, *Middle East International,* 10 January 2003. Scott Sagan and Kenneth Waltz, *The Spread of Nuclear Weapons* (Norton, 1995), pp. 18–19.

12. Robert S. Greenberger, *Wall Street Journal,* 21 March 2003.

13. *Ha'aretz* and *Jerusalem Post,* 4 December 2002. United Nations Security Council Resolution 252 (21 May 1968).

14. Steven R. Weisman, *New York Times,* 15 March 2003. Text of the president's address, *New York Times,* 15 March 2003.

15. Noam Chomsky interviewed by Cynthia Peters, ZNet, 9 March 2003.

16. Rachel Meeropol, ed., *America's Disappeared* (Seven Stories Press, 2005).

2. COLLATERAL LANGUAGE

1. Randal Marlin, *Propaganda and the Ethics of Persuasion* (Broadview Press, 2002), p. 66.

2. For references and more detail, see Noam Chomsky, *Necessary Illusions* (South End Press, 1989), pp. 16–17.

3. Michael Dawson, *The Consumer Trap* (University of Illinois Press, 2003).

4. Stuart Ewen, *Captains of Consciousness* (McGraw-Hill, 1976), p. 85.

5. Both quotes are from the Federal Convention of 1787. See Rufus King, *Life and Correspondence of Rufus King* (G. P. Putnam's Sons, 1894), vol. 1, pp. 587–619; and Robert Yates, "Notes of the Secret Debates of the Federal Convention of 1787," from *Documents Illustrative of the Formation of the Union of the American States* (Government Printing Office, 1927).

6. Harold Lasswell, "Propaganda," *Encyclopedia of the Social Sciences* (Macmillan, 1935), pp. 521–28.

7. Adam Nagourney and Richard W. Stevenson, *New York Times*, 5 April 2003.

8. Martin Sieff, *American Conservative*, 4 November 2002.

9. Howard LaFranchi, *Christian Science Monitor*, 14 January 2003. Linda Feldmann, *Christian Science Monitor*, 14 March 2003. Jim Rutenberg and Robin Toner, *New York Times*, 22 March 2003.

10. On the impact of sanctions, see Anthony Arnove, ed., *Iraq Under Siege*, 2nd ed. (South End Press, 2002). See also Carl Kay-

sen et al., *War with Iraq* (American Academy of Arts and Sciences, Committee on International Security Studies, 2002).

11. Department of State, *World Military Expenditures and Arms Transfers* (WMEAT), 6 February 2003.

12. Ruth Leacock, *Requiem for Revolution* (Kent State University Press, 1990), p. 33.

13. Executive Order 12513, Prohibiting Trade and Certain Other Transactions Involving Nicaragua. See also *New York Times*, 2 May 1985; and Noam Chomsky, *Turning the Tide* (South End Press, 1986), p. 144, for more detail.

14. Jim Rutenberg, *New York Times*, 1 April 2003.

15. Charles Glass, *London Review of Books*, 17 April 2003.

16. Neely Tucker, *Washington Post*, 3 December 2002. Neil A. Lewis, *New York Times*, 9 January 2003.

17. Jack M. Balkin, *Los Angeles Times*, 13 February 2003. See also Rachel Meeropol, ed., *America's Disappeared* (Seven Stories Press, 2005).

18. Winston Churchill cited by A. W. Brian Simpson, *Human Rights and the End of Empire* (Oxford University Press, 2001), p. 55.

19. *Nightline* special edition, ABC News, 31 March 2003.

20. David Lloyd George cited by V. G. Kiernan, *European Empires from Conquest to Collapse, 1815–1960* (Leicester University Press/Fontana Paperbacks, 1982), p. 200.

21. Kate Zernike, *New York Times*, 5 April 2003.

3. REGIME CHANGE

1. Editorial, *New York Times*, 6 August 1954.

2. State Department Policy Planning Council (1964) cited in Piero Gleijeses, *Conflicting Missions* (University of North Carolina Press, 2002), p. 26.

3. The Research Unit for Political Economy, *Monthly Review* 55, no. 1 (May 2003).

4. William Stivers, *Supremacy and Oil* (Cornell University Press, 1982), pp. 28–29, 34; Stivers, *America's Confrontation with Revolutionary Change in the Middle East* (St. Martin's Press, 1986), pp. 20ff.

5. Graphic accompanying the article by James Dao and Eric Schmitt, *New York Times*, 7 May 2003.

6. Jawaharlal Nehru, *The Discovery of India* (Asia Publishing House, 1961), p. 326. For discussion, see Noam Chomsky, *Towards a New Cold War* (New Press, 2003), p. 228.

7. Woodrow Wilson's minister of the interior cited in Gordon Connell-Smith, *The Inter-American System* (Oxford University Press, 1966), p. 16.

8. Selig Harrison et al., *Turning Point in Korea* (Report of the Task Force on U.S. Korea Policy) (Center for International Policy/The Center for East Asian Studies, University of Chicago, 1 March 2003).

9. Zbigniew Brzezinski, *The Grand Chessboard* (Basic Books, 1998), p. 40.

10. Joseph A. Schumpeter, *Imperialism and Social Classes,* ed. Paul Sweezy (A. M. Kelly, 1951), p. 68.

11. Editorial, *Monthly Review* 54, no. 7 (December 2002).

12. William A. Williams, *Empire as a Way of Life* (Oxford University Press, 1982).

13. For a discussion of the background, see Noam Chomsky, *Deterring Democracy,* expanded ed. (Hill and Wang, 1992), pp. 47–49.

14. Michael Ignatieff, *New York Times Magazine,* 5 January 2003. See also Ignatieff, *New York Times,* 28 July 2002; and Ignatieff, *Empire Lite* (Penguin, 2003).

15. John Stuart Mill, "A Few Words on Non-Intervention" (1859), in Mill, *Collected Works,* vol. 21 (University of Toronto Press, 1984), pp. 109–24.

16. Michael Ignatieff, *New York Times Magazine,* 7 September 2003. See also Noam Chomsky, *Rogue States* (South End Press, 2000).

17. Samuel Huntington, *Foreign Affairs* 78, no. 2 (March–April 1999).

18. "Japan Envisions a 'New Order' in Asia, 1938," reprinted in Dennis Merrill and Thomas G. Paterson, eds., *Major Problems in American Foreign Relations,* 5th ed., vol. 2 (Houghton Mifflin, 2000). See also Noam Chomsky, *American Power and the New Mandarins* (Pantheon, 1969).

19. Antonio Gramsci quoted by Vincente Navarro, *The Politics of Health Policy* (Blackwell, 1994), p. 1.

4. WARS OF AGGRESSION

1. Errol Morris, director, *The Fog of War* (Sony Pictures Classics, 2003).

2. For Taylor's account of the standards at Nuremberg, see Telford Taylor, *Nuremberg and Vietnam* (Quandrangle, 1970), pp. 37–38; and Taylor, *The Anatomy of the Nuremberg Trials* (Knopf, 1992), pp. 398ff.

3. A. Frank Reel, *The Case of General Yamashita* (University of Chicago Press, 1949), p. 174.

4. G. John Ikenberry, *Foreign Affairs* 81, no. 5 (September–October 2002).

5. Madeleine K. Albright, *Foreign Affairs* 82, no. 5 (September–October 2003).

6. Henry A. Kissinger, *Chicago Tribune*, 11 August 2002.

7. George W. Bush, Remarks by the President on Iraq, Cincinnati Museum Center, Cincinnati, Ohio, 7 October 2002.

8. Tim Weiner, *New York Times*, 9 May 2005. See also discussion and references in Noam Chomsky, *Hegemony or Survival* (Owl Books, 2004), pp. 86–87.

9. Duncan Campbell, *The Guardian* (London), 7 April 2003. Catherine Wilson, Associated Press, 10 March 2004.

10. Juan Forero, *New York Times*, 29 January 2004.

11. Julian Borger, *The Guardian* (London), 17 April 2002. Rupert Cornwell, *The Independent* (London), 17 April 2002. Katty Kay, *The Times* (London), 17 April 2002.

12. Jason B. Johnson, *San Francisco Chronicle*, 24 January 2005. Daniel Grann, *Atlantic Monthly* 287, no. 6 (June 2001). Leslie Casmir, *Daily News* (New York), 14 December 2000.

13. Testimony of Robert Jackson, 21 November 1945, in *Trial of the Major War Criminals before the International Military Tribunal*, vol. 2 (International Military Tribunal, 1947).

14. Testimony of Sir Hartley Shawcross, 4 December 1945, in *Trial of the Major War Criminals before the International Military Tribunal*, vol. 2.

15. Taylor, *The Anatomy of the Nuremberg Trials*.

16. For further discussion, see Noam Chomsky, *Fateful Triangle*, rev. ed. (South End Press, 1999), chaps. 5, 9.

17. Jacques Lanusse-Cazale and Lorna Chacon, Agence France-Presse, 3 November 2003.

18. Paul Lewis, *New York Times*, 24 December 1989 and 30 December 1989.

19. For further discussion, see Noam Chomsky, *Deterring Democracy*, expanded ed. (Hill and Wang, 1992).

20. Michael J. Glennon, *Foreign Affairs* 82, no. 3 (May–June 2003); and *Foreign Affairs* 78, no. 3 (May–June 1999).

21. Carsten Stahn, *American Journal of International Law* 97, no. 4 (October 2003).

22. See, among others, Oxford Research International Poll, December 2003; Guy Dinmore, *Financial Times* (London), 11 September 2003; and Patrick E. Tyler, *New York Times*, 24 September 2003.

23. Walter Pincus, *Washington Post*, 12 November 2003.

24. William Stivers, *Supremacy and Oil* (Cornell University Press, 1982), pp. 28–29, 34.

25. Thom Shanker and Eric Schmitt, *New York Times*, 20 April 2003. Stephen Barr, *Washington Post*, 29 February 2004. Walter Pincus, *Washington Post*, 23 January 2004. John Burns and Thom Shanker, *New York Times*, 26 March 2004.

26. Allan Beattie and Charles Clover, *Financial Times*, 22 September 2003. Jeff Madrick, *New York Times*, 2 October 2003. Thomas Crampton, *New York Times*, 14 October 2003.

27. Madrick, *New York Times*, 2 October 2003. George Anders and Susan Warren, *Wall Street Journal*, 19 January 2004.

28. Robert McNamara, *In Retrospect* (Times Books, 1995). For a full discussion, see Noam Chomsky, Z, July–August 1995.

29. Mohamed El-Baradei, *New York Times*, 12 February 2004.

30. General Lee Butler, National Press Club, Washington, D.C., 2 February 1998.

31. *Ha'aretz* (Hebrew edition), 10 February 2004.

32. Air Force Space Command, "Strategic Master Plan (SMP) FY04 and Beyond," 5 November 2002.

33. See William Arkin, *Los Angeles Times*, 14 July 2002; Julian Borger, *The Guardian* (London), 1 July 2003; and Michael Sniffen, Associated Press, 1 July 2003.

34. William J. Broad, *New York Times*, 1 May 2000.

35. Scott Peterson, *Christian Science Monitor,* 6 May 2004. David Pugliese, *Ottawa Citizen,* 11 January 2001.

36. Peter Schwartz and Doug Randall, *An Abrupt Climate Change Scenario and Its Implications for United States National Security* (October 2003). Report commissioned by the U.S. Defense Department.

37. Robert Repetto and Jonathan Lash, *Foreign Policy,* no. 108 (fall 1997).

38. John Vidal, *The Guardian* (London), 16 February 1996. Thomas Land, *Toronto Star,* 30 March 1996. See also reports of the International Panel on Climate Change (IPCC).

39. Hannah Arendt, *Eichmann in Jerusalem* (Penguin, 1994).

40. McGeorge Bundy, *Danger and Survival* (Random House, 1988), p. 326.

5. HISTORY AND MEMORY

1. Frank Diaz Escalet, *Obispo Romero y los Martires-Jesuitas de El Salvador [Bishop Romero and the Jesuit Martyrs of El Salvador]* (1995). Original painting in the Organization of the American States Museum, Washington, D.C.

2. Marjorie Hyer, *Washington Post,* 4 April 1980.

3. Larry Rohter, *New York Times,* 10 September 1989.

4. Lindsey Gruson, *New York Times,* 17 November 1989. The murdered Jesuit priests were Ignacio Ellacuria Beas Coechea, Ignacio Martin-Baro, Segundo Montes Mozo, Amando Lopez

Quintana, Juan Ramon Moreno, and Joaquin Lopez y Lopez. The Jesuits' cook, Julia Elba Ramos, and her daughter, Celina, were also murdered. For more discussion, see Noam Chomsky, *Deterring Democracy*, expanded ed. (Hill and Wang, 1992).

5. Carla Anne Robbins, *Wall Street Journal*, 27 April 2004.

6. William Safire, *New York Times*, 22 April 1985.

7. R. W. Apple, Jr., *New York Times*, 11 June 2004.

8. Robert Pear, *New York Times*, 14 January 1989.

9. John M. Goshko, *Washington Post*, 26 October 1983.

10. Joanne Omang, *Washington Post*, 2 May 1985. For the full text of the Executive Order, see *New York Times*, 2 May 1985.

11. Lou Cannon and Joanne Omang, *Washington Post*, 4 March 1986.

12. Transcript of President Reagan's speech, *New York Times*, 28 October 1983. See Stuart Taylor, Jr., *New York Times*, 6 November 1983, for acknowledgment of some of the many distortions in the case for attacking Grenada.

13. Francis X. Clines, *New York Times*, 13 December 1983.

14. Alan Pertman, *Boston Globe*, 15 July 1992.

15. Elisabeth Bumiller and Elizabeth Becker, *New York Times*, 8 June 2004.

16. Elizabeth Becker, *New York Times*, 27 May 2004.

17. Noam Chomsky, *At War With Asia* (Pantheon, 1970; AK Press, 2004), p. 223.

18. Christine Hauser, *New York Times*, 14 April 2004.

19. See National Security Archive Electronic Briefing Book No. 4, online at http://www.gwu.edu/~nsarchiv/NSAEBB/NSAEBB4/.

20. Peter Smith, *Talons of the Eagle* (Oxford University Press, 1996), p. 137.

21. Stephen Schlesinger and Stephen Kinzer, *Bitter Fruit*, updated ed. (Harvard University Press, 1999).

22. Stephen Schlesinger, *The Nation* 265, no. 2 (14 July 1997).

23. See Piero Gleijeses, *Politics and Culture in Guatemala* (University of Michigan Press, 1988).

24. Peter Grier, *Christian Science Monitor*, 7 May 1984. Douglass Farah, *Washington Post*, 11 March 1999.

25. Tim Weiner, *New York Times*, 7 June 1997.

26. Thomas McCann, *An American Company* (Crown, 1976), p. 47.

27. Eqbal Ahmad, *Terrorism: Theirs and Ours* (Seven Stories Press, 2002).

28. Werner Daum, *Harvard International Review* 23, no. 2 (summer 2001). Jonathan Belke, *Boston Globe*, 22 August 1999.

29. Eqbal Ahmad, *Confronting Empire* (South End Press, 2000), p. 135.

30. Jason Burke, *Al-Qaeda* (I. B. Tauris, 2004).

31. Richard Clarke, *Against All Enemies* (The Free Press, 2004).

32. Burke, *Al-Qaeda*, p. 239.

33. Barry Schweid, Associated Press, 11 June 2004.

34. Max Boot, *Financial Times* (London), 17 June 2004.

35. Sebastian Rotella, *Los Angeles Times*, 4 November 2002. Jimmy Burns and Mark Huband, *Financial Times* (London), 24 January 2003. Eric Lichtblau, *New York Times*, 25 January 2003. Marlise Simons, *New York Times*, 29 January 2003. Philip Shenon, *New York Times*, 4 March 2003.

6. THE DOCTRINE OF GOOD INTENTIONS

1. Philip Stephens, *Financial Times* (London), 19 November 2004.

2. Sam Allis, *Boston Globe*, 29 April 2004.

3. David Ignatius, *Washington Post*, 2 November 2003.

4. Patrick E. Tyler, *New York Times*, 1 April 2003. Dexter Filkins, *New York Times*, 1 April 2003. Tyler Hicks and John F. Burns, *New York Times*, 3 April 2003. Robert Collier, *San Francisco Chronicle*, 3 April 2003.

5. Noam Chomsky, *Deterring Democracy*, expanded ed. (Hill and Wang, 1992), p. viii.

6. Clive Ponting, *Winston Churchill* (Sinclair-Stevenson Ltd., 1994), p. 132.

7. Noam Chomsky, *At War With Asia* (Pantheon, 1970; AK Press, 2004).

8. John K. Fairbank, presidential address, American Historical Association annual meeting, New York, New York, December 29, 1968, published in the *American Historical Review* 74, no. 3 (February 1969).

9. See Noam Chomsky and Edward S. Herman, *Manufacturing Consent*, 2nd ed. (Pantheon, 2002), p. 173.

10. John F. Burns, *New York Times*, 29 November 2004.

11. Bernard Fall, *Last Reflections on a War* (Doubleday, 1967).

12. Howard Kurtz, *Reliable Sources*, CNN, 22 August 2004.

13. Richard A. Oppel, Jr., Robert F. Worth et al., *New York Times*, 8 November 2004. Photograph by Shawn Baldwin.

14. Richard A. Oppel, Jr., *New York Times*, 8 November 2004.

15. Protocol Additional to the Geneva Conventions of 12 August 1949, and relating to the Protection of Victims of Non-International Armed Conflicts, 8 June 1977, Part III, "Wounded, Sick, and Shipwrecked." See also Dahr Jamail, "Iraqi Hospitals Ailing Under Occupation," 21 June 2005, online at http://dahrjamailiraq.com.

16. U.S. War Crimes Act of 1996 (18 U.S.C. 2441).

17. Les Roberts et. al, *The Lancet* 364, no. 9448 (20 November 2004). See also the comment on the report by Richard Horton, *The Lancet* 364, no. 9448.

18. Patrick Wintour and Richard Norton-Taylor, *The Guardian* (London), 30 October 2004.

19. Sarah Boseley, *The Guardian* (London), 11 March 2005. Rory McCarthy, *The Guardian* (London), 9 December 2004.

20. Justin Lewis, Sut Jhally, and Michael Morgan, "The Gulf War: A Study of the Media, Public Opinion and Public Knowledge," Center for the Study of Communication, Department of Communication, University of Massachusetts at Amherst (February 1991).

21. Hatfield Consultants (Vancouver), *Development of Impact Mitigation Strategies Related to the Use of Agent Orange Herbicide in the Aluoi Valley, Viet Nam* (2000) and *Preliminary Assessment of Environmental Impacts Related to Spraying of Agent Orange Herbicide During the Viet Nam War* (1998). Reuters, *Boston Globe*, 7 March 2002. Associated Press, *Taipei Times*, August 2003.

22. Barbara Crossette, *New York Times*, 18 August 1992.

23. Doug Struck, *Washington Post*, 18 April 2001. Colin Joyce, *Daily Telegraph* (London), 21 April 2001. David McNeill, *New Statesman*, 26 February 2001.

24. Rory McCarthy, *The Guardian* (London), 15 November 2004. Steve Negus, *Financial Times* (London), 12 November 2004.

25. Michael Janofsky, *New York Times*, 13 November 2004.

26. Eric Schmitt, *New York Times*, 17 November 2004.

27. Michael D. Sallah, Mitch Weiss, and Joe Mahr, *Toledo Blade*, 22 October 2003–5 September 2004.

28. Fall, *Last Reflections on a War*.

29. Chomsky, *At War With Asia*.

30. Noam Chomsky, *New York Review of Books* 13, no. 12 (1 January 1970), reprinted in Chomsky, *At War With Asia*.

31. See *Manufacturing Consent*, directors Mark Achbar and Peter Wintonick (Zeitgeist Films, 1993), and the accompanying book of the same title published by Black Rose Books in Montréal in 1994.

32. See David Cortright, *Soldiers in Revolt*, updated ed. (Haymarket Books, 2005).

33. For further discussion of this topic, see Noam Chomsky, *Understanding Power*, ed. Peter R. Mitchell and John Schoeffel (New Press, 2002), chap. 7, note 57.

34. Chicago Council on Foreign Relations, "American Public Opinion and Foreign Policy," *Global Views 2004;* and polls from the Program on International Policy Attitudes (PIPA), University of Maryland.

35. Bryan Bender, *Boston Globe*, 7 October 2004. Demetri Sevastopulo, *Financial Times* (London), 27 April 2005.

36. PIPA, "Bush Supporters Still Believe Iraq Had WMD or Major Program, Supported al Qaeda," 21 October 2004. Howard LaFranchi, *Christian Science Monitor*, 22 October 2004. Bob Herbert, *New York Times*, 10 September 2004. Robert P. Laurence, *San Diego Union Tribune*, 14 October 2003.

37. Chicago Council on Foreign Relations, *Global Views 2004*, p. 14.

38. Gardiner Harris, *New York Times*, 31 October 2004.

39. Fareed Zakaria, *Newsweek*, 11 October 2004.

7. INTELLECTUAL SELF-DEFENSE

1. *BBC World News,* 3 December 2004.

2. Thomas E. Ricks, *Washington Post,* 9 May 2004.

3. PIPA/Knowledge Networks Poll, Press Release, 3 December 2003; and additional PIPA polls.

4. Edmund L. Andrews, *New York Times,* 3 December 2004.

5. Adam Smith, *An Inquiry into the Nature and Causes of the Wealth of Nations* (1776) (University of Chicago Press, 1996), book 4, chap. 2.

6. David Ricardo, *The Principles of Political Economy and Taxation* (Dover, 2004), pp. 83–84.

7. Lord Hutton, "Report of the Inquiry into the Circumstances Surrounding the Death of Dr. David Kelly C.M.G.," 28 January 2004.

8. Noam Chomsky, *Necessary Illusions* (South End Press, 1989), p. viii.

9. David Hume, *Of the First Principles of Government* (Longmanns, Green, and Company, 1882), chap. 1.

10. *KidsPost, Washington Post,* 12 November 2004.

11. See Howard Zinn, *SNCC,* updated ed. (South End Press, 2002); and Zinn, *You Can't Be Neutral on a Moving Train,* updated ed. (Beacon, 2002).

12. Ralph Atkins et al., *Financial Times,* 22 November 2004.

13. For details, see Roger Morris, *New York Times,* 14 March 2003; and Saïd K. Aburish, *Saddam Hussein* (Bloomsbury, 2000).

14. Reginald Dale, *Financial Times,* 1 March 1982. See also Reginald Dale, *Financial Times,* 28 November 1984.

15. Thomas L. Friedman, *New York Times,* 14 May 2003.

16. See Anthony Arnove, ed., *Iraq Under Siege,* updated ed. (South End Press, 2002); and John Mueller and Karl Mueller, *Foreign Affairs* 78, no. 3 (May–June 1999).

17. Les Roberts et al., *The Lancet* 364, no. 9448 (20 November 2004). See also the comment on the report by Richard Horton, *The Lancet* 364, no. 9448.

18. H. Bruce Franklin, *War Stars* (Oxford University Press, 1988).

19. Lyndon Johnson, *Congressional Record,* 15 March 1948, House of Representatives, 80th Congress, 2nd Session, vol. 94, part II (Government Printing Office, 1948), p. 2883.

20. Lyndon Johnson, Remarks to American and Korean Servicemen at Camp Stanley, Korea, 1 November 1966, *Public Papers of the Presidents, 1966,* Book II (Government Printing Office, 1967), p. 253.

21. Noam Chomsky, *Hegemony or Survival* (Owl Books, 2004), pp. 1–2, 236–37.

22. John Steinbruner and Nancy Gallagher, *Dædalus* 133, no. 3. (summer 2004)

8. DEMOCRACY AND EDUCATION

1. David Barsamian and Noam Chomsky, *Propaganda and the Public Mind* (South End Press, 2001), p. 19.

2. Jeffrey Dubner, *The American Prospect* (April 2005).

3. Kathy Lynn Gray, *Columbus Dispatch*, 27 January 2005, quoting Ohio Republican senator Larry A. Mumper.

9. ANOTHER WORLD IS POSSIBLE

1. John Lewis Gaddis, *Surprise, Security, and the American Experience* (Harvard University Press, 2004). John Quincy Adams, letter to George Erving, 29 November 1818, in Worthington Chauncey Ford, ed., *Writings of John Quincy Adams* (Macmillan, 1916), p. 483.

2. Joy Olson and Adam Isacson, *Just the Facts* (Latin America Working Group, 1998–2001).

3. Raymond Hernandez and Al Baker, *New York Times*, 9 January 2005. Mike Allen and Peter Baker, *Washington Post*, 7 February 2005.

4. Steffie Woolhandler, Terry Campbell, and David U. Himmelstein, *International Journal of Health Services* 34, no. 1 (2004); and David U. Himmelstein, Steffie Woolhandler, and Sidney M. Wolfe, *International Journal of Health Services* 34, no. 1 (2004).

5. See, among others, the National Public Radio/Kaiser/Kennedy School poll, 5 June 2002.

6. David K. Shipler, *Los Angeles Times*, 6 March 2005.

7. Stephen Barr, *Washington Post*, 30 October 2003.

INDEX

INDEX

ABOUT THE AUTHORS

NOAM CHOMSKY is the author of numerous bestselling political works, from *American Power and the New Mandarins* in the 1960s to *Hegemony or Survival* in 2003. A professor of Linguistics and Philosophy at MIT, he is widely credited with having revolutionized modern linguistics. He lives outside Boston, Massachusetts.

DAVID BARSAMIAN, founder and director of an award-winning and widely syndicated weekly show, *Alternative Radio* (www.alternativeradio.org) has authored several books of interviews with leading political thinkers, including Arundhati Roy, Howard Zinn, Edward Said, and especially Noam Chomsky. He lives in Boulder, Colorado.

NOAM CHOMSKY

HEGEMONY OR SURVIVAL

From the world's foremost intellectual activist, an irrefutable analysis of America's pursuit of total domination and the catastrophic consequences that are sure to follow.

The United States is in the process of staking out not just the globe, but the last unarmed spot in our neighbourhood – the skies – as a militarized sphere of influence. Out earth and its skies are, for the Bush administration, the final frontiers of imperial control. In *Hegemony or Survival*, Noam Chomsky explains how we came to this moment, what kind of peril we find ourselves in, and why our rulers are willing to jeopardize the future of our species. With the striking logic that is his trademark, Chomsky dissects America's quest for global supremacy, tracking the US government's aggressive pursuit of policies intended to achieve 'full spectrum dominance' at any cost. Laying out the rules of militarization of space, the ballistic-missile defence program, unilateralism, the Iraqi crisis and the dismantling of international agreements, he argues that, in our era, empire is a recipe for an earthly wasteland.

'Chomsky is one of the most significant challengers of unjust power and delusions; he goes against every assumption about American altruism and humanitarianism' Edward W. Said

'A superb polemicist who combines fluency of language with a formidable intellect' *Observer*